Second Edition

Exam-Oriented
BIOCHEMISTRY

S Rajan MSc, PhD

Assistant Professor
Department of Microbiology
MR Government Arts College
Mannargudi

R Selvichristy MSc, MPhil

Microbiologist

CBS

CBS Publishers and Distributors Pvt Ltd

New Delhi • Bengaluru • Chennai • Kochi • Kolkata • Mumbai
Bhopal • Bhubaneswar • Hyderabad • Jharkhand • Nagpur • Patna
Pune • Uttarakhand • Dhaka (Bangladesh) • Kathmandu (Nepal)

Second Edition

Exam-Oriented

BIOCHEMISTRY

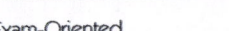

ISBN: 978-93-89396-28-7

Second Edition: 2020
First Edition: 2017

Published by Satish Kumar Jain and produced by Varun Jain for

CBS Publishers and Distributors Pvt Ltd

4819/XI Prahlad Street, 24 Ansari Road, Daryaganj, New Delhi 110 002, India.
Ph: 23289259, 23266861, 23266867   Website: www.cbspd.com
Fax: 011-23243014   e-mail: delhi@cbspd.com; cbspubs@airtelmail.in.
Corporate Office: 204 FIE, Industrial Area, Patparganj, Delhi 110 092
Ph: 4934 4934   Fax: 4934 4935   e-mail: publishing@cbspd.com;
publicity@cbspd.com

Branches

- **Bengaluru:** Seema House 2975, 17th Cross, K.R. Road, Banasankari 2nd Stage, Bengaluru 560 070, Karnataka
  Ph: +91-80-26771678/79   Fax: +91-80-26771680   e-mail: bangalore@cbspd.com
- **Chennai:** 7, Subbaraya Street, Shenoy Nagar, Chennai 600 030, Tamil Nadu
  Ph: +91-44-26680620, 26681266   Fax: +91-44-42032115   e-mail: chennai@cbspd.com
- **Kochi:** 42/1325, 1326, Power House Road, Opp KSEB, Power House, Ernakulam 682 018, Kochi, Kerala
  Ph: +91-484-4059061-65   Fax: +91-484-4059065   e-mail: kochi@cbspd.com
- **Kolkata:** 6/B, Ground Floor, Rameswar Shaw Road, Kolkata-700 014, West Bengal
  Ph: +91-33-22891126, 22891127, 22891128   e-mail: kolkata@cbspd.com
- **Mumbai:** 83-C, Dr E Moses Road, Worli, Mumbai-400018, Maharashtra
  Ph: +91-22-24902340/41   Fax: +91-22-24902342   e-mail: mumbai@cbspd.com

Representatives

Bhopal	0-8319310552	Bhubaneswar	0-9911037372	Hyderabad	0-9885175004
Jharkhand	0-9811541605	Nagpur	0-9421945513	Patna	0-9334159340
Pune	0-9623451994	Uttarakhand	0-9716462459	Dhaka	01912-003485
Kathmandu	977-9818742655			(Bangladesh)	
(Nepal)					

Printed at: City Printers, Delhi, India

Preface

Biochemistry, the branch of science concerned with the chemical and physico-chemical processes and substances which occur within living organisms. This is a rapidly progressing field with multiple applications. We are teaching microbiology and biochemistry to undergraduate and postgraduate courses in microbiology degree. I understand the students of microbiology are struggling to complete allied biochemistry. Hence we are intended to prepare a book for biochemistry for microbiology and biotechnology students in the name of exam companion in Biochemistry. This book covers almost all basic chapters of Biochemistry include carbohydrates, proteins, lipids, vitamins, nucleic acids, minerals, pigments, blood, hormones, enzymes, metabolism of organic matters, pH, Buffers, electrophoresis, centrifugation, etc. This book is prepared for the benefit of students for their university examinations. Question bank and practical part also included for the benefit of students. Salient features of this book include Simple language, Understandable Concepts, Proper illustrations, Student Exam friendly approach, Low cost, Practical, Unit wise Coverage and Question Bank. We are sure, this book will bring good motivation among the students. This book will also improve the skill of the students. We welcome suggestions from readers for improving the contents of the book. Please forward your comments to ksrajan99@gmail.com.

S Rajan
R Selvichristy

Contents

CHAPTER 1

CARBOHYDRATES

Definition

Carbohydrates are defined as polyhydroxy aldehydes or ketones. Substance that gives polyhydroxy aldehydes or ketones on hydrolysis is called as carbohydrates. Carbohydrates are hydrates of carbon. It contains carbon, hydrogen and oxygen in the ratio of 1:2:1. It contains carbonyl group (Aldose or keto) and atleast two hydroxyl groups. Glyceroldehyde and dihydroxy acetone are the simplest carbohydrates. Carbon dioxide and water are combined to form carbohydrate with the liberation of oxygen. General formula for carbohydrate is $Cn(H_2O)_n$.

$$CO_2 + H_2O \longrightarrow (CH_2O) + O_2\uparrow$$
Carbohydrate

Classification

Carbohydrates are classified into two based on sweetness. They are sugars and nonsugars. Sugars are sweet in taste. Eg., Monosaccharides and Oligosaccharides. Non sugars are formed by linking monosaccharides. These are sweetless carbohydrates. Eg. Polysaccharides. They are insoluble in water (**Fig.1**).

Carbohydrates are generally classified into 3 major groups based on the number of monomers. They are as follows.

1. Monosaccharides   2. Oligosaccharides and   3. Polysaccharides.

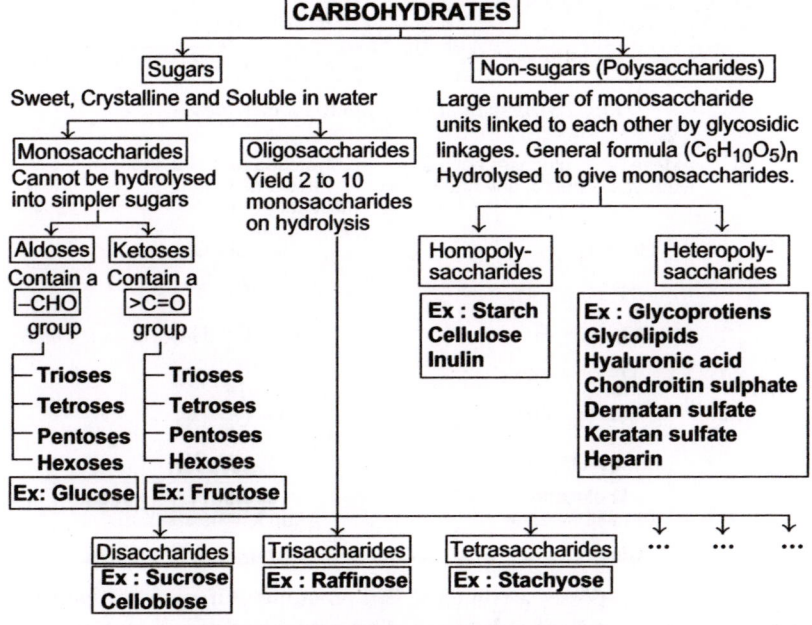

Fig. 1 : *Classification of Carbohydrates.*

MONOSACCHARIDES

These are simple carbohydrates. They are called as simple sugar. Monosaccharides cannot be hydrolysed into more simpler form. The general formula for monosaccharide is $Cn(H_2O)_n$. They are sweet in taste. They are soluble in water. They are crystalline in nature. They contain 3-10 carbon atoms, 2 or more hydroxyl groups and one aldehyde (CHO) or keto (CO) group. Monosaccharides reduce oxidizing agents such as hydrogen peroxide. Depending upon the number of carbon atoms monosaccharides are subdivided into trioses, tetroses, pentoses, hexoses and heptoses (**Fig.2**). Monosaccharides are divided into two groups according to their functional groups (**Fig.3**). They are Aldoses (CHO group) and Ketoses (CO group). Aldoses are sugars containing aldehyde group eg : glucose, galactose, mannose. Ketoses are sugars containing ketone group eg : fructose and sorbose (**Fig.2 and Fig.3**).

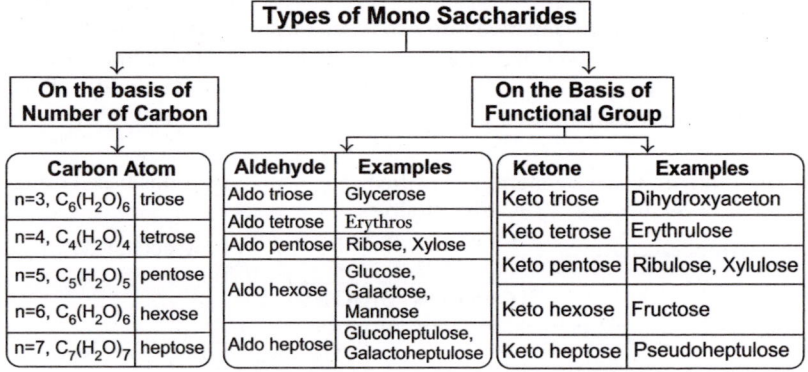

Carbon Atom		**Aldehyde**	**Examples**	**Ketone**	**Examples**
n=3, $C_6(H_2O)_6$	triose	Aldo triose	Glycerose	Keto triose	Dihydroxyaceton
n=4, $C_4(H_2O)_4$	tetrose	Aldo tetrose	Erythros	Keto tetrose	Erythrulose
		Aldo pentose	Ribose, Xylose		
n=5, $C_5(H_2O)_5$	pentose		Glucose,	Keto pentose	Ribulose, Xylulose
		Aldo hexose	Galactose,		
n=6, $C_6(H_2O)_6$	hexose		Mannose	Keto hexose	Fructose
n=7, $C_7(H_2O)_7$	heptose	Aldo heptose	Glucoheptulose, Galactoheptulose	Keto heptose	Pseudoheptulose

Fig. 2 : *Classification of Monosaccharides.*

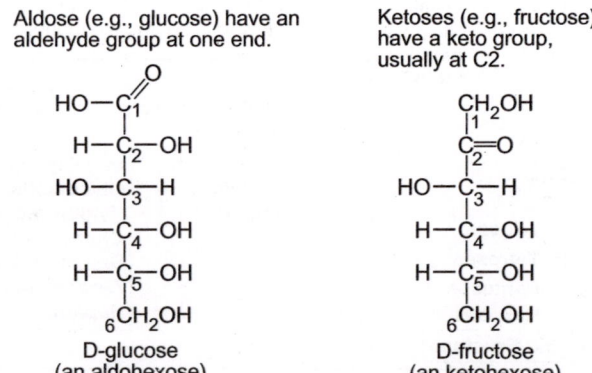

Fig. 3 : *Classification of Monosaccharides–Functional groups.*

Carbohydrates posses asymmetric carbon atoms. An asymmetric carbon atom is a carbon atom that is attached to four different types of atoms or groups of atoms. This is also called chiral carbon (**Fig.4**).

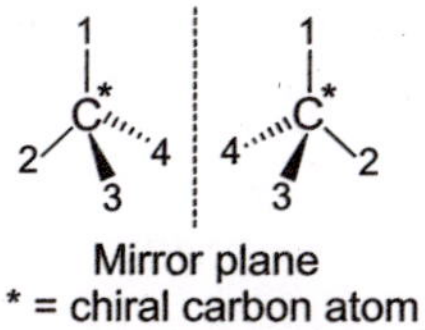

Mirror plane
* = chiral carbon atom

Fig. 4 : *Asymmetric Carbon*

Simplest form of aldose – Glyceraldehyde; Simplest form of ketose – Dihydroxy acetone; Commonest aldose – Glucose; Commonest ketose – Fructose.

Trioses

It is a simple form of monosaccharide. It contains three carbon atoms. Eg. Glyceraldehyde. It is a simple sugar. They are sweet in taste. They are soluble in water. They are crystalline in nature. It contains alddhyde group or keto group. Triose containing aldo group are called aldotrioses. Eg. Glycerose or glyceraldehyde Triose containing keto group are called ketotrioses or triulose. Eg. Dihydroxyacetone (**Fig.5**).

D-Glyceraldehyde   Dihydroxyacetone

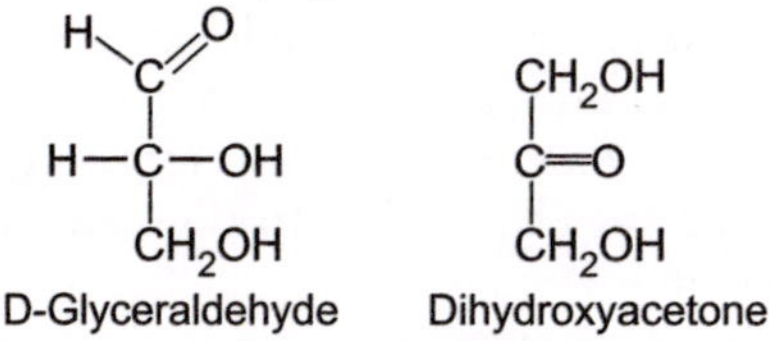

Fig. 5 : *Trioses.*

Tetroses

It is one of the monosaccharide. It contains four carbon atoms. Eg. Erythrose. It is a simple sugar. They are sweet in taste. They are soluble in water. They are crystalline in nature. It contains alddhyde group or keto group (**Fig.6**). Tetrose containing aldo group are called aldotetroses. Eg. Erythrose. Tetrose containing keto group are called ketotetroses. Eg. Erythrulose. Xylulose is a metabolite of glucuronic acid.

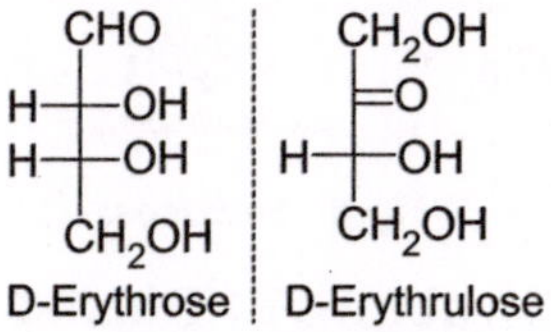

D-Erythrose ┊ D-Erythrulose

Fig. 6 : *Tetroses.*

Pentoses

It is one of the monosaccharide. It contains Five carbon atoms. Eg. Ribose. ($C_5H_{10}O_5$). It is a simple sugar. They are sweet in taste. They are soluble in water. They are crystalline in nature. It contains aldehyde group or keto group. Pentose containing aldo group are called aldopentose. Eg. Ribose. Pentose containing keto group are called ketopentoses or pentulose. Eg. Ribulose. Ribose sugar occurs in RNA. Deoxyribose sugar exists in DNA. The methypentose fucose occurs in human milk as well as in blood. Pentasons are found in would gums. Ribulose is an important component of photosynthesis (**Fig.7**).

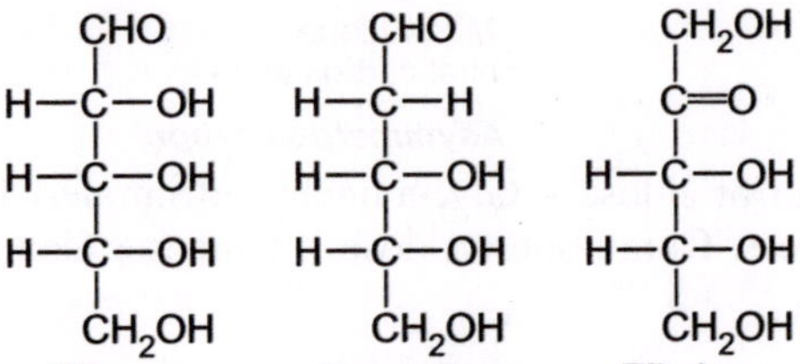

```
      CHO              CHO              CH₂OH
  H—C—OH          H—C—H            C=O
  H—C—OH          H—C—OH        H—C—OH
  H—C—OH          H—C—OH        H—C—OH
      CH₂OH            CH₂OH            CH₂OH
   Ribose          Deoxyribose        Ribulose
```

*Fig. 7 : **Pentoses.***

Hexoses

Hexoses are monosaccharides containing 6 carbon atoms. The molecular formula of hexose is $C_6H_{12}O_6$. It is a simple sugar. They are sweet in taste. They are soluble in water. They are crystalline in nature. It contains alddhyde group or keto group. Hexose containing aldo group are called aldohexoses. Eg. Glucose. Hexose containing keto group are called ketohexoses. Eg. Fructose. Aldohexoses contain asymmetric carbon atoms at position 2,3,4 and 5. Hence an aldohexose can exist in 16 isomeric forms. ($2n = 24 = 16$). The ketohexoses contain 3 asymmetric carbon atoms at position 3,4 and 5. Hence, it exist in 8 isomeric forms ($2n = 23 = 8$). Hexoses are biologically important sugars. Eg. Glucose, Fructose, galactose etc., (**Fig.8**).

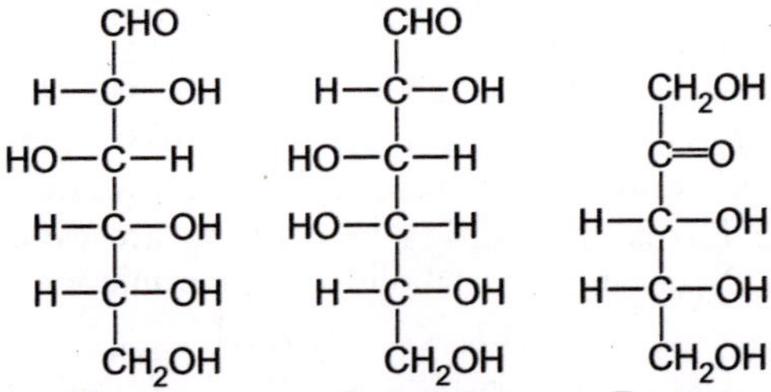

```
      CHO              CHO              CH₂OH
  H—C—OH          H—C—OH           C=O
 HO—C—H          HO—C—H        H—C—OH
  H—C—OH         HO—C—H        H—C—OH
  H—C—OH          H—C—OH
      CH₂OH            CH₂OH            CH₂OH
   Glucose          Galactose         Fructose
```

*Fig.8 : **Hexoses.***

Structure of Glucose

Glucose is a simple sugar. It is a monosaccharide. It cannot be hydrolysed further. They are sweet in taste. They are soluble in water. They are crystalline in nature. It is sparingly soluble in alcohol. It melts at 146°C. D–Glucose is also called **dextrose**. Human blood contains 60-100 mg of glucose in 100 ml of blood in fasting. It serves as the major metabolic fuel in cells and tissues. Oxidation of glucose quickly provides energy for the cells. Hence, glucose is described as the chief source of energy. In 1890, Fitting and Baeyer proposed the straight chain structure for glucose. It doesnot showed any mutarotation.

In this structure, the 6 carbon atoms of glucose are arranged in a straight chain. It is also called open chain structure.

$$CH_2OH—CHOH—CHOH—CHOH—CHOH—CHO$$

Bayers structure is associated with an aldehyde group and five hydroxyl groups. Later **Fischer** proposed another straight chain projection formula to explain its sterioisomers. This straight chain structure explains the existence of 4 asymmetric carbon at position 2,3,4 and 5 (**Fig.9**). Hence an aldohexose can exist in 16 isomeric forms (2n = 24 = 16). Glucose in open chain structure does not show mutarotation. In order to explain these properties, which are not consistent with the open chain structure, a ring structure was proposed for glucose.

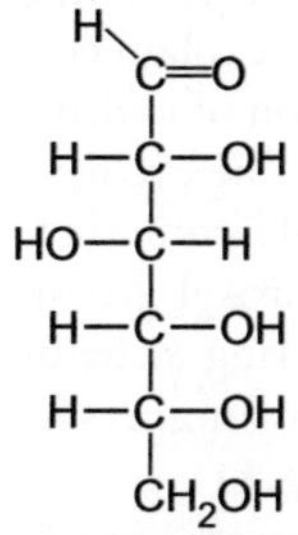

Fig. 9 : *Open Structure of Glucose.*

Cyclic structure of glucose : Ring structure of Glucose explain the properties which are not explained by open chain structure because Ring structure has no free aldehydic group, glucose does not respond to certain characteristic tests of aldehydes, like Schiff's test and addition reaction with sodium-bisulphite. The cyclic structure is attributed to the formation of hemiacetal the cyclic structure thus formed is a six-membered ring. This cyclic structure has one chiral centre more than the open chain structure. Therefore, two possible isomeric forms are possible. They are α-D Glucopyranose and β-D-Glucopyranose (**Fig.10**).

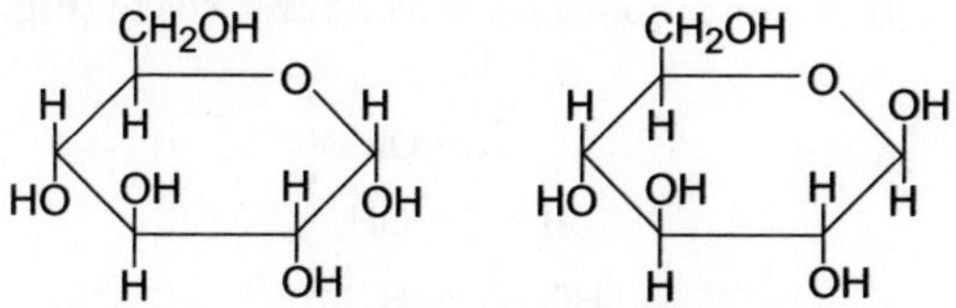

α-D-(+)-Glucopyranose   β-D-(+)-Glucopyranose

Fig. 10 : *Cyclic Structure of Glucose.*

Haworth structures

In cyclic structure, the atoms are arranged in the form of a ring. Howorth in 1929 devised this ring structure. Hence this ring structure is called Howort projection formula. The six-membered ring structures of glucose can be called pyranose structures, in analogy with pyran. Hence, the anomers are called α-D-(+)-glucopyranose and β-D-(+)-glucopyranose. According to Howorth two pyranose ring structures are proposed. They are alpha D plus gluco pyranose and beta D plus gluco pyranose.

FRUCTOSE

It is a simple sugar. It is a monosaccharide. It cannot be hydrolysed further. They are sweet in taste. They are soluble in water. They are crystalline in nature. It occurs in plants and honey. It is a reducing sugar. It is an optically active compound. It plays an important role in cellular metabolism. It rotates lights in left direction. So it is called levo sugar.

Structure of Fructose

Fructose is a ketone and consists of six carbon atoms in a straight chain with the keto functional group at position 2 of the carbon chain (**Fig.8**). As the open chain structure fails to explain certain facts like its existence in two isomeric forms and the formation of hydrogen sulphite addition product, the ring structure was established. The cyclic structure is a five-membered ring. Therefore, there are two possible isomeric forms. The two cyclic forms differ in the configuration of the hydroxyl group at C2. These isomers are called anomers. The five-membered ring structure of fructose is called furanose. The cyclic structures of the two anomers are named α-D-(-)-fructofuranose and β-D-(-) fructofuranose (**Fig.11**).

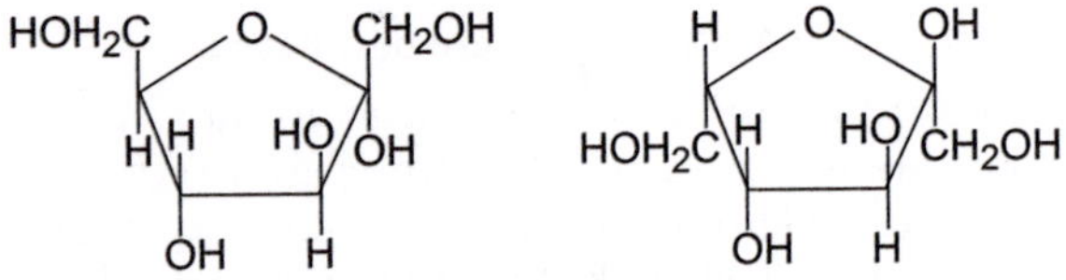

Fig. 11 : **Cyclic Structure of Fructose.**

Galactose : It is a monosaccharide. It is a simple sugar. It is a component of milk sugar. In liver, galactose is changed into glucose. It is a reducing sugar. It is soluble in water. It is sweet in taste. It is an aldohexose. It is in crystalline form. It occurs in molasses, malted milk shake (**Fig.12**).

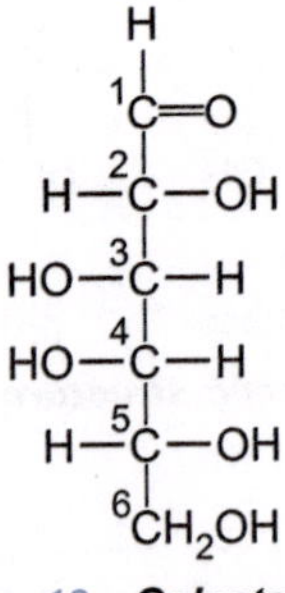

Fig. 12 : **Galactose.**

Ribose : It is a monosaccharide. It is a pentosugar. Molecular formula is $C_5H_{10}O_5$. It is a simple sugar. It is a reducing sugar. It is soluble in water. It is sweet in taste. It is in crystalline form. It exhibits isomerism. It remains in cyclic and straight chain structure. It is an aldo pentose. It is a major component of RNA. It occurs in all plants and animals (**Fig.13**).

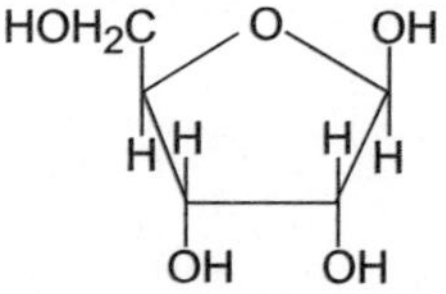

Fig. 13 : *Ribose.*

Properties of monosaccharides

1. Colour and shape : Monosaccharides are colourless and crystalline compounds.

2. Solubility : They are readily soluble in water.

3. Taste : They have sweet taste.

4. **Stereo isomerism** : Compounds with same molecular formula but differ in their configuration are called stereoisomers. Asymmetric carbon allows the formation of streoisomerism. Types of streoisomerism are D and L isomerism, Optical isomerism, Epimerism and anomerism.

(a) D and L isomerism : It is a mirror image of each other. These two forms are called enantiomers. For example glucose can exist in two forms. They are L form and D-Form (**Fig.14**). D-glucose and L-glucose are mirror images of each other.

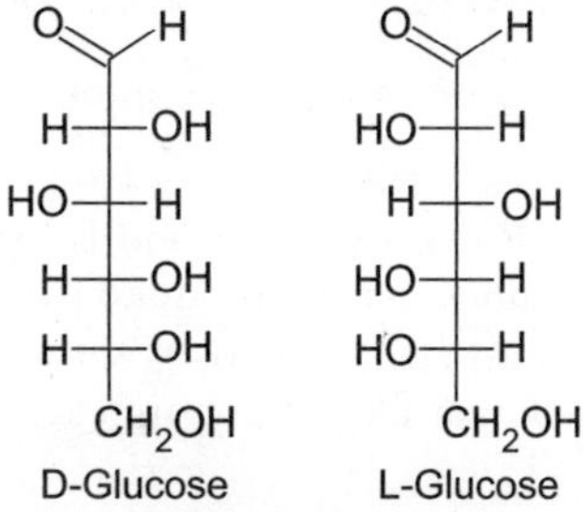

Fig. 14 : *D and L Form.*

(b) Optical Isomerism : Optical activity is the capasity of a substance to rotate plane polarised light is passing through it. When light is rotate to the left, it is described as levorotation. when plane polarised light rotates to the right, it is described as dextrorotation. This phenomenon exhibited by asymmetric compounds is called optical isomerism (**Fig. 14**).

(c) Epimerism : Epimers are sugars which are differ with each other with respect to single carbon, other than anomeric carbon. Galactose and maltose are the epimers of glucose. They differ from glucose with respect to the C4 and C2 respectively (**Fig.15**).

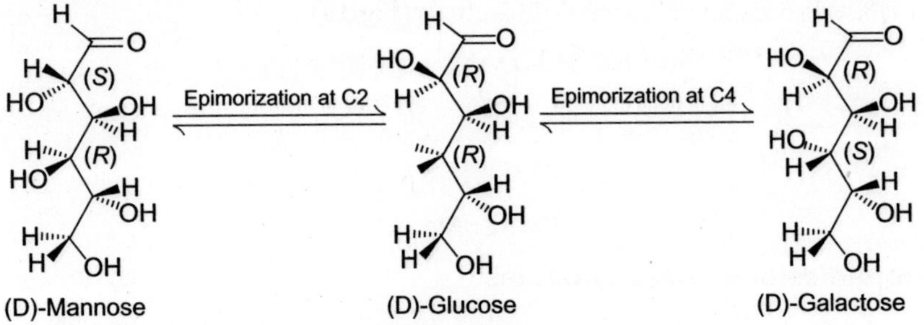

Fig. 15 : *Epimers.*

(d)　Anomerism : In solution glucose predominantly exists as closed chain structure. Because of cyclization of sugar, an additional asymmetric center is created at C1 (anomeric Carbon). This leads to formation of of two isomers namely α-D-Glucopyranose and β-D-Glucopyranose (**Fig.16**).

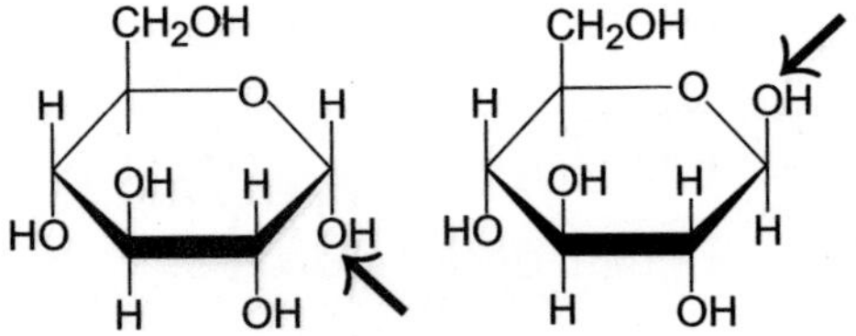

Fig. 16 : *Anomeric Carbon.*

5. Mutarotation : It is defined as the change in the specific optical rotation representing interconversion of α and β forms of D glucose to an equilibrium mixture. The mutarotation is due to the existence of two optical isomers of glucose, namely α,D-glucose with a specific rotation at +112.2° and β,D-glucose with a specific rotation at +18.7° (**Fig. 17**).

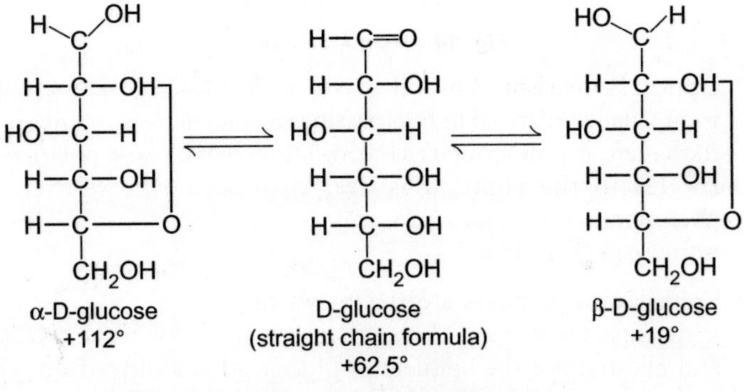

Fig. 17 : *Muta rotation.*

6. Glucoside formation :

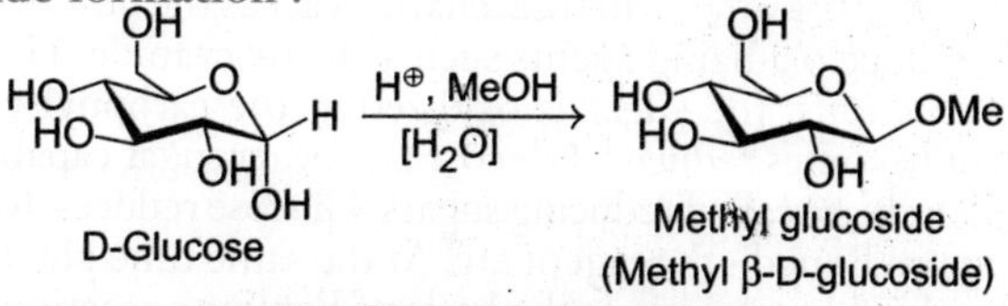

D-Glucose

Methyl glucoside
(Methyl β-D-glucoside)

Fig. 18 : *Glycoside formation.*

Glucose reacts with methanol in the presence of HCl and gives α and β glucoside. Glucoside formation is due to the reaction of alcohol with glucoside –OH group of glucose.α,D glucose is formsβ,D-methyl glucoside. In the same way, fructose forms fructoside (**Fig. 18**).

7. Oxidation : Glucose when treated with bromine water, forms gluconic acid. The aldehyde group is oxidised to carboxylic group. First, bromine forms hypobromous acid (HOBr), with water and oxidises the glucose to gluconic acid. When glucose is oxidised with nitric acid, saccharic acid is formed (**Fig.19**).

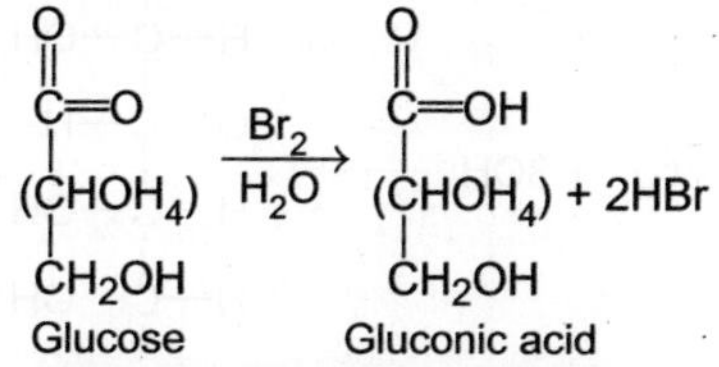

Glucose   Gluconic acid
Fig. 19 : *Oxidation of Glucose.*

8. Reduction : Monosaccharides can be reduced by various reducing agents such as sodium amalgam or by hydrogen under high pressure in the presence of catalysts. The reduction is due to the presence of aldehyde or ketone group. On reduction they yield alcohols. When glucose is reduced by sodium amalgam, sorbitol is formed. Mannose yields mannitol and fructose yeilds a mixture of sorbitol and mannitol because of the formation of new asymmetric carbon C2 of fructose.

9. Reaction with concentrated H_2SO_4 : Glucose is treated with concentrated H_2SO_4 or HCl, and forms 5, hydroxymethyl furfural which on further heating yields levulinic acid and formic acid (**Fig. 20**). This reaction is the basis of the colour test, known as **Molish test** for sugars. When pentoses are treated with mineral acids furfural is obtained on heating.

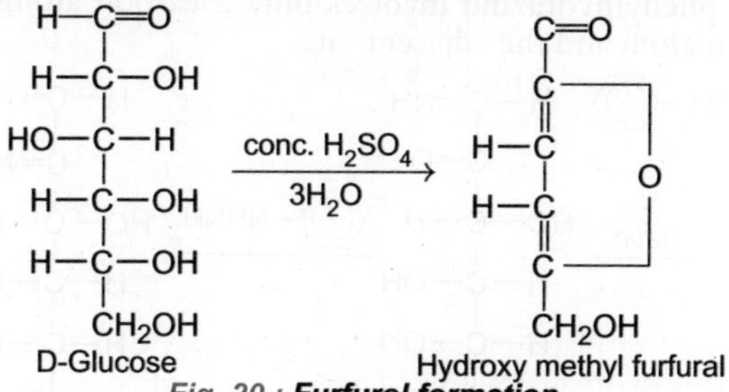

D-Glucose   Hydroxy methyl furfural
Fig. 20 : *Furfurol formation.*

10. Ester formation : Monosaccharides can form esters with carboxylic acids due to the presence of OH groups. For eg. glucose reacts with five molecules of acetic anhydride to form penta acetate derivative. It obviously indicates that the glucose contain five OH groups.

11. Reducing property : Monosaccharides act as the best reducing agents. They readily reduce oxidizing agents such as ferric cyanide, H_2O_2 and cupric ion. In such reactions, the sugar is oxidized at the carbonyl group and the oxidising agent becomes reduced. Glucose and other sugar capable of reducing certain compounds are called reducing sugars. Glucose reduces Tollen's reagent, Fehling's reagent, Benedict's reagent etc. At the same time glucose is oxidized to gluconic acid. This property is the basis of Fehling's reaction (ammoniacal cupric sulphate), a qualitative test for the presence of reducing sugar. In this reaction, Cu^{2+} is reduced into Cu^+ and at the same time glucose is oxidised to gluconic acid. During this reaction the blue colour of the reagent changes to reddish orange colour. A standard test for the presence of reducing sugar is the reduction of Ag+ in ammonia solution (Tollen's reagent) to yield a metallic silver mirror lining on the sides of the test tube (**Fig. 21**).

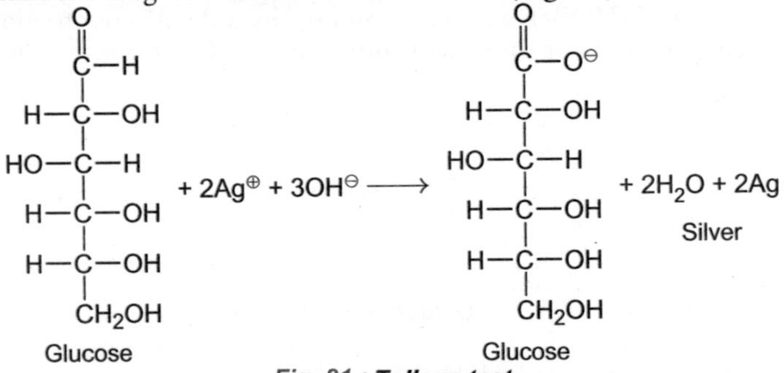

Fig. 21 : *Tollens test.*

12. Reaction with alanine : The aldehyde group of glucose condenses with the amino group of alanine to form **Schiff's base.** Fructose also gives Schiff's base with alanine. The browning reaction occurs during baking of bread and other mixtures of carbohydrates and proteins is believed to be due to the formation of Schiff's base between the amino groups of proteins and the aldehyde groups of carbohydrates.

13. Ozazone formation : An important reaction of reducing sugars, (monosaccharide and disaccharides) having potential aldehyde or ketone group, is their action with phenylhydrazine to form phenyl hydrazones. Reaction with phenylhydrazine involves only 2 carbon atoms namely the carbonyl carbon atom and the adjacent one.

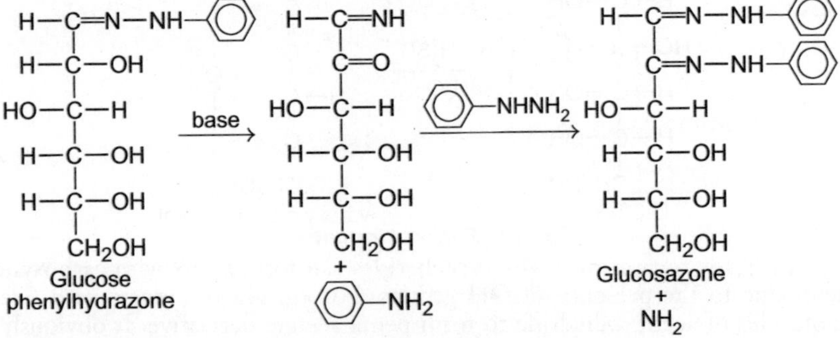

Fig. 22 : *Ozazone formation.*

The steps involved in phenylhydrazine reactions are

1. One molecule of glucose condenses with one molecule of phenyl hydrazine to form soluble glucose phenylhydrazone.

2. In the presence of excess of phenylhydrazine, another molecule of phenylhydrazine enters the reaction.

3. Now the third molecule of phenylhydrazine enters the reaction, giving rise to phenyl glucosazone which are yellow coloured crystals. The shape of the crystals and the time of formation of osazone differ for various sugars.

The reaction of phenyl hydrazine with fructose is similar to glucose. Here again 3 molecules of phenyl hydrazine take part in the reaction. Fructose gives fructosazone. Disaccharides such as maltose and lactose also exhibit the property of osazone formation. But, sucrose does not form osazone, since it does not contain free CHO or CO groups which are responsible for the reducing property.

14. **Fermentation** : Fermentation is the process of converting a larger complex molecule into simple molecules by means of enzymes in an anaerobic condition. The products of the reaction are alcohol and CO_2.

$$C_6H_{12}O_6 \xrightarrow[\text{Catalyst}]{\text{Zymase}} 2C_2H_5OH + 2CO_2$$

Fructose/Glucose   **Ethanol**

Functions of Monosaccharides

1. Ribose is a structural element of nucleic acids and also of some coenzymes.

2. Glucose on oxidation yield energy which is required for various metabolic activities.

3. Fructose is found in fruits, honey etc.which are responsible for sweetness and can be converted to glucose and utilised in the body.

4. Galactose is a component of milk sugar-lactose, glycolipids and glycoproteins

5. Mannose is a constituent of mucoproteins and glycoproteins which are essential for the body.

Occurrence of monosaccharide

- Ribose is a constituent of nucleic acid.
- Occurs in polysaccharides like starch.
- Gum arabic, cherry gums, wood gums, proteoglycans.
- Fruit juices and cane sugar contains monosaccharides
- Plant mannosans and glycoproteins are the best source of monosaccharides.
- Intermediate in carbohydrate metabolism.

OLIGOSACCHARIDES

These are carbohydrates that yield two to ten molecules of same or different types of monosaccharides on hydrolysis. They have sweet in taste. They are soluble in water. Oligosaccharides are one of the components of fibre, found in plants. The general formula is $C_n(H_2O)_n$-1 (Eg) Lactose, Maltose and

Sucrose. The monosaccharide units are united by a glycosidic linkage. Oligosaccharides are formed by the condensation of two molecules of monosaccharides with the elimination of one molecule of water. When oligosaccharides containing two molecules of monosaccharides, the oligosaccharides are called disaccharides. Trisaccharide contain three units of monosaccharides **(Fig.23)**.

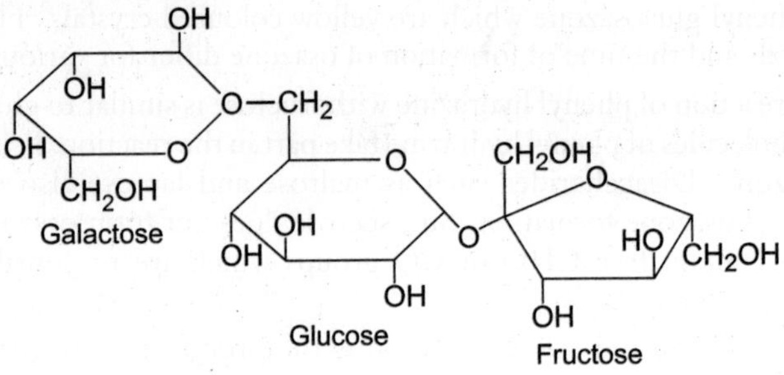

Fig. 23 : **Raffinose.**

Classification of oligosaccharides

Oligosaccharides are classified into three or more types based on the number of monosaccharide molecule released on hydrolysis. They are

Disaccharides, Trisaccharides, Tetrasaccharides

Disaccharides

It yields 2 molecules of monosaccharides on hydrolysis. In disaccharides, monosaccharides are linked by the glycosidic bonds. The properties of the disaccharides depend to a great extent on the type of linkage. If the two potential aldehyde or ketone of both monosaccharides are involved in the linkage, the sugar will not exhibit reducing properties and will not be able to form osazones. eg. Sucrose. But if one of them is not bound in this way, it will permit reduction and osazone formation by the sugars eg. lactose and maltose which are known as reducing disaccharides.

Maltose

It is a disaccharide. It is commonly called malt sugar. Maltose is the end product of digestion of starch by the action of salivary amylase, in the mouth and pancreatic amylase in the intestine. Maltose is composed of two glucose molecules combined by α-1,4 glycosidic linkage. Malt from sprouting barley is the major source of maltose. It is a rather sweet sugar and is highly soluble in water. The structure of maltose shows that the potential aldehyde group of glucose -2 is blocked in the glycosidic linkage, whereas the potential aldehyde group of glucose is free and can reduce alkaline copper solution. It is because of this free aldehyde in the first glucose molecule, maltose has reducing property. Maltose is formed as an intermediate product in the intestine. Maltose is split into two molecules of glucose by the enzyme **maltase** of the intestinal juice before absorption **(Fig.24)**.

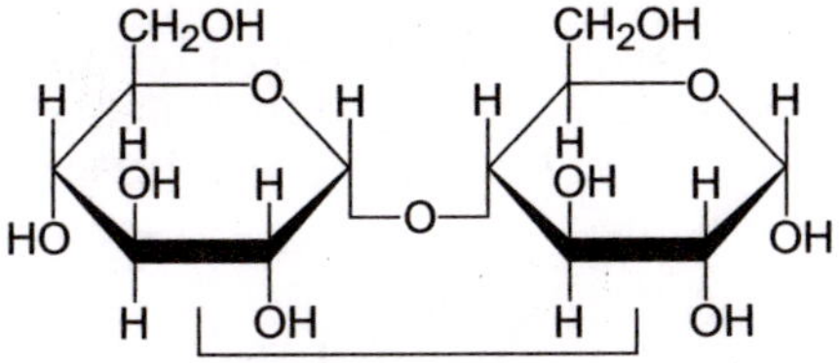

Fig. 24 : **Maltose.**

Lactose

It is a disaccharide. It is of animal origin. Lactose is commonly called as milk sugar. It is present in the milk of mammals. Lactose is found in the urine of pregnant and lactating women. It is less soluble in water and less sweeter than sucrose. Lactose is composed of α-D-Galactose and α-D-Glucose held together by β-(1,4)-glycosidic linkage between glucose and galactose molecules. The anomeric carbon of C1 glucose is free, hence lactose exhibits reducing (**Fig.25**). Lactose properties. Lactose form ozazones. It also reduces Fehling's solution. Lactose is hydrolysed by acids or by the enzyme **lactase.** The intestine of milk sucking infants has the enzyme lactase, which converts lactose into glucose and galactose. Then only it is absorbed in the body. Excess of lactose ingested into the body causes diarrhoea, abnormal intestinal flow and colic pain. Lactose is not fermented by yeast.

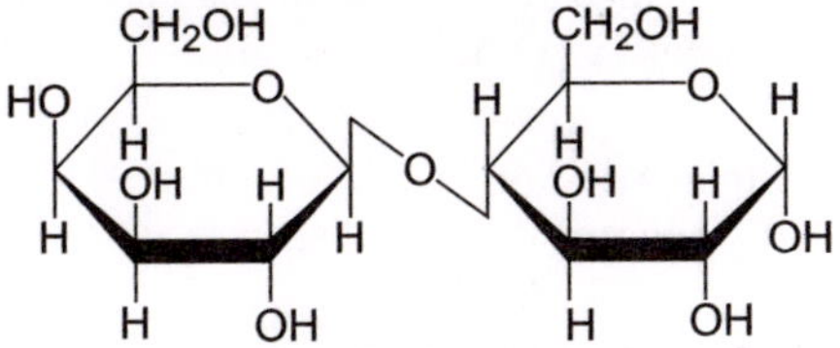

Fig. 25 : **Lactose.**

Sucrose

It is a disaccharide. Sucrose is called as *"table sugar"*. It is also called as *"cane sugar"* as it can be obtained from sugar cane. The molecular formula of sucrose is $C_{12}H_{22}O_{11}$. In sucrose fructose and glucose rings are linked to each other by an oxygen atom. Sucrose typically forms crystals. It is widely distributed in sugar cane, beet root, pine apple, honey, carrot and ripe fruits. Sucrose consists of one molecule of glucose and one molecule of fructose. The linkage between these molecules is formed between the aldehyde group of glucose and the ketone group of fructose. Thus, both the potential aldehyde group of glucose and the ketone group of fructose are blocked in the linkage and sucrose has no free reducing group. On account of this structural peculiarity sucrose is a non-reducing sugar. It does not reduce Tollen's and Fehling's solutions and does not form osazone. Sucrose on hydrolysis by dilute acids or the enzyme sucrase or invertase gives a mixture of glucose and fructose. It is called as invert sugar. Sucrose is dextrorotatory (+62.5°) but it's hydrolytic products are levorotatory because fructose has a greater specific levo-rotation than the dextrorotation of glucose. As the hydrolytic products invert the rotation, sucrose is known as invert sugar and the process is called as invertion. Honey contains plenty of *'invert sugar'* and the presence of fructose accounts for the greater sweetness of honey. Hydrolysis of sucrose by invertase liberates glucose and fructose (**Fig.26**).

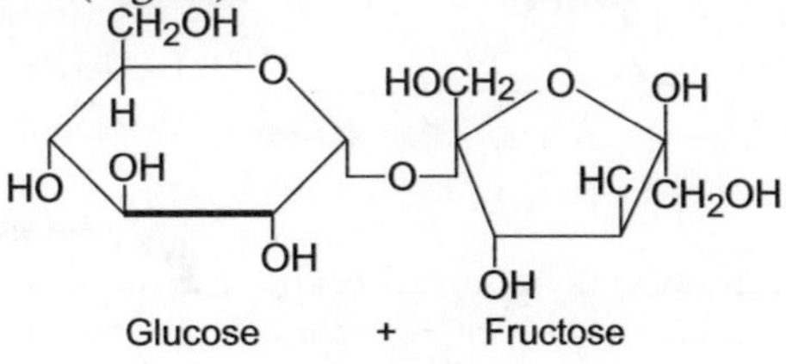

Glucose   +   Fructose

Sucrose

Fig. 26 : Sucrose.

Cellobiose

It is a disaccharide. It is identical with maltose. Cellobiose is an isomer of maltose obtained by partial hydrolysis of cellulose. It gives two molecules of glucose on hydrolysis. Maltose and Cellobiose are disaccharide made up of 2 glucose molecules, and both are linked together between C-1 of one sugar and C-4 of the other sugar. But the "first" glucose in maltose is the alpha anomer, and the "first" glucose in cellobiose is the beta anomer.

Reducing sugars bear a free aldehyde or keto group. They have the capacity to reduce cupric ions to cuprous ions. Eg., Maltose, melibiose, cellobiose.

Non reducing sugars do not have free aldehyde or keto group. They fail to reduce cupric ions to cuprous ions. Eg., Sucrose, trehalose.

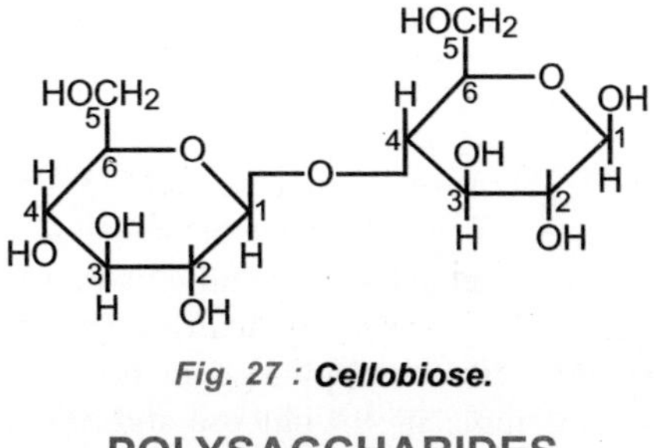

Fig. 27 : Cellobiose.

POLYSACCHARIDES

They are non sugars. It yields more than 10 monosaccharide units on hydrolysis. They are not have sweet taste. They are amorphous substances insoluble in water. They do not exhibit any of the properties of aldehyde and keto groups. Monosaccharides are linked together by glycosidic bonds in polysaccharides. Polysaccharides, which are also known as glycans composed of number of monosccharide units. They represent condensation products of several molecules of simple sugars or monosaccharides. They form linear chain or branched chain molecules. They contain only one type of monosaccharide units or many types of monosaccharide units. According to this nature polysaccharides are classified into two groups, homopolysaccharides and heteropolysaccharides.

Homopolysaccharides

Homopolysaccharides are composed of only one type of monosaccharides. On hydrolysis they yield only one type of monosaccharides Eg. starch, glycogen, cellulose, inulin, pectin and hemicelluloses. They yield only glucose on hydrolysis.

Starch

It is a homopolysaccharide. It is a non sugar. It yields glucose on complete hydrolysis. This is the reserve food in plants. They are abundantly found in root, stem, vegetables, fruits and cereals. Rice, wheat and vegetables are a rich source of starch. Starch occurs in the forms of grain which may be spherical or oval in shape. Microscopically, the starch grains are found to differ in size and shape according to their sources. Starch is made up of two structurally different homopolysaccharide units. They are amylose and amylopectin. Glucose molecules are arranged in a linear form in amylose where as glucose molecules are arranged in a highly branched form in amylopectin. Amylose has (1-4)-glycosidic linkages. The glycosidic bond occurs between OH group of C1 in one glucose unit is joined to that of C4 of the next unit. A solution of starch react with iodine to give blue colour. The blue colour formation is mainly due to the presence of amylose in starch.

Amylose : It has a simple unbranched structure . It is soluble in water. It has $\alpha,(1-4)$-glycosidic linkages. Gives blue colour with dilute iodine solution. The molecular weight ranges from 10,000-50,000

Amylopectin : It has branched chain structure. It is insoluble in water, can absorb water and swells up. It has $\alpha,(1-4)$-glycosidic and $\alpha,1-6$ glycosidic linkages. It gives yellow or orange colour with iodine solution. The molecular weight ranges from 50,000 – 1,00,000.

Starch digestion

Digestion of starch begins in the mouth. Salivary enzyme α amylase randomly hydrolyses all the $\alpha,1-4$ glycosidic bonds of starch. Starch digestion continues in the small intestine under the influence of pancreatic amylase. This enzyme degrades starch to a mixture of disaccharide maltose and the trisaccharide maltotriose. These oligosaccharides are hydrolysed to their component monosaccharides by specific enzymes persent in the brush border membranes of the intestinal mucosa. The resulting monosaccharides are absorbed in the intestine and transported to the blood stream.

Glycogen

Glycogen is a homopolysaccharide. It gives only glucose units on hydrolysis. It is the major reserve carbohydrate in animals. It is also called animal starch. It is stored in liver. Glycogen is present in all cells of skeletal muscle and liver and occur as cytoplasmic granules. Glycogen is readily available as immediate source of energy. During starvation glycogen is mobilised from the storage tissue and converted to glucose by the enzyme glycogen phosphorylase. Formation of glycogen from glucose is called as **Glycogenesis** and breakdown of glycogen to form glucose is called as **Glycogenolysis**. It gives red colour with iodine. It is non reducing sugar. It is a white powder readily dissolved in water.

Cellulose

Cellulose is an organic compound with the formula $(C_6H_{10}O_5)n$. It is a polysaccharide consisting of a linear chain of several hundred to many thousands of linked D glucose units. Cellulose is an important structural component of the primary cell wall of green plants. The cellulose content of cotton fiber is 90%, that of wood is 40–50% and that of dried hemp is approximately 57%. Cellulose has no taste, is odourless. It is insoluble in water and most organic solvents. It is biodegradable. It can be broken down chemically into its glucose units by treating it with concentrated mineral acids at high temperature. Cellulose is derived from D-glucose units, which condense through -glycosidic bonds. Cellulose is a straight chain polymer. Plant-derived cellulose is usually found in a mixture with hemicellulose, lignin, pectin .

Chitin

Chitin is a modified polysaccharide that contains nitrogen. It is synthesized from units of N-acetyl-D-glucosamine. These units form covalent β-(1→4)-linkages. Chitosan is produced commercially by deacetylation of chitin. Chitosan is soluble in water. Chitin is not soluble in water. Humans and other mammals have chitinase and chitinase-like proteins that can degrade chitin. They also possess several immune receptors that can recognize chitin and its degradation products in a pathogen-associated molecular pattern, initiating an immune response

Inulin

Inulins are a group of naturally occurring polysaccharides. It is produced by many types of plants. The inulins belong to a class of dietary fibres known as fructans. Inulin is used by some plants as a means of storing energy and is typically found in roots or rhizomes. Most plants that synthesize and store inulin do not store other forms of carbohydrate such as starch. Inulin is a heterogeneous collection of fructose polymers. It consists of chain-terminating glucosyl moieties and a repetitive fructosyl moiety which are linked by β (2,1) bonds. Inulin is not digested by enzymes in the human alimentary system, contributing to its functional properties: reduced calorie value, dietary fiber and prebiotic effects. Without colour and odour, it has little impact on sensory characteristics of food products. Nonhydrolyzed inulin can also be directly converted to ethanol in a simultaneous saccharification and fermentation process, which may have great potential for converting crops high in inulin into ethanol for fuel. Inulin enhances the growth and activities of bacteria or inhibits growth or activities of certain pathogenic bacteria. Inulins are polymers composed mainly of fructose units, and typically have a terminal glucose. In general, plant inulins contain between 20 and several thousand fructose units. Smaller compounds are called fructo oligosaccharides, the simplest being 1-kestose, which has 2-fructose units and 1-glucose unit.

HETEROPOLYSACCHARIDES : Heteropolysaccharides is a mixture of different types of monosaccharides. On hydrolysis, they yield a mixture of monosaccharides. Eg. Hyaluronic acid, Heparin. Hyaluronic acid is made up of glucuronic acid and N-acetyl glucosamine They are classified into neutral sugars and mucopolysaccharides.

Mucopolysaccharides (agar agar) : The heteropolysaccharides situated in extra cellular matrix are called as mucopolysaccharides. (eg). Hyaluronic acid, heparin, keratan sulphate and chondroitin sulphate.

Neutral sugars : It give morethan one type of sugar units on hydrolysis eg. Hemicelluloses, mucillages and pectic substances.

Hemicellulose : A **hemicelluloses is also called as** polyose. It is a group of heterogeneous polysaccharides. Hemicellulose function as supporting material in the cell wall. The amount of hemicellulose of the dry weight of wood is usually between 20 and 30%. Mannose is the most important hemicellulosic monomer followed by xylose, glucose, galactose and arabinose. It is easily hydrolyzed by dilute acid or base as well as myriad hemicellulase enzymes.Hemicelluloses are synthesised from sugar nucleotides in the cell's Golgi apparatus.

Mucopolysaccharides : Mucopolysaccharides are long chains of sugar molecules that are found throughout the body, often in mucus and in fluid around the joints. They are more commonly called glycosaminoglycans. These are gelatinous substances. They are heteropolysaccharides. Molecular weight is upto 5 million. These are extracellular materials. Examples are Hyaluronic acid, Heparin, Agar agar, Chondroitin.

Hyaluranic acid

It is a heteropolysaccharide. It is a mucopolysaccharide. Hyaluronic acid is also called hyaluronan. It is an anionic, nonsulfatedglycosaminoglycan distributed widely throughout connective, epithelial, and neural tissues. It is unique among glycosaminoglycans in that it is nonsulfated, forms in the plasma membrane instead of the Golgi apparatus, and can be very large, with its molecular weight often reaching the millions. It acts as a lubricant and as a biological cement in connective tissue. It is a straight chain polymer.

Chondroitin

It is a heteropolysaccharide. It is a mucopolysaccharide. It is found in cartilages. It is a parent substance for chondroitin sulfate. They are straight chain polymer linked by β-(1-4)-glycosidic bond.

Heparin

It is a heteropolysaccharide. It is a mucopolysaccharide. It is used as an anticoagulant (blood thinner). It contains uronic acid and sulphuric acid. Hence it is called as acidic heteropolysaccharide.

Agar agar

It is a heteropolysaccharide. It is a mucopolysaccharide. It is used in biology laboratory as solidifying agent. Agar-agar is a jelly-like substance obtained from algae. Agar is derived from the polysaccharide agarose, which forms the supporting structure in the cell walls of certain species of algae, and which is released on boiling. These algae are known as agarophytes and belong to the phylum Rhodophyta (red algae). Agar is actually the resulting mixture of two components: the linear polysaccharide agarose, and a heterogeneous mixture of smaller molecules called agaropectin. Agar can be used as a laxative, an appetite suppressant, a vegetarian substitute for gelatin. The gelling agent in agar is an unbranched polysaccharide obtained from the cell walls of some species of red algae, primarily from the genera Gelidium and Gracilaria. For commercial purposes, it is derived primarily from Gelidium amansii. Agar is a polymer made up of subunits of the sugar galactose.

Functions of carbohydrates

1. They supply energy for body functions and for doing work.

2. They are structural components of many organisms.

3. They exert a sparing action on proteins.

4. They provide the carbon skeleton for the synthesis of some nonessential amino acids and fats.

5. Some carbohydrates are present as tissue constituents.

6. Starch forms main source of carbohydrates in the diet.

7. Glycogen is the major carbohydrate reserve in animals and is often called animal starch. It is stored in liver and muscle of animals.

8. Cellulose is widely distributed in plant sources. It occurs in the cell walls of plants where it contributes to the structure. It is the main consituent of the supporting tissues of plants and forms a considerable part of vegetables.

9. Pectin and hemicellulose are present in fruits of many plants and serve as jelling agents.

10. Hyaluronic acid occurs in synovial fluid, in skin and in tissues. It acts as a cementing substance in tissues and also acts as a lubricant. It is also present in vitreous humor.

11. Heparin is used in medicine as an anticoagulant and prevents blood clotting.

12. Keratan sulphate is an important component of cartilage and cornea.

13. Lactose is otherwise called as milk sugar. It is present in milk and is made up of monosaccharides—glucose and galactose.

Glucose + Galactose → Lactose

14. Maltose is otherwise known as 'malt sugar' and is present in germinating cereals, malt etc.It is the intermediate product in the hydrolysis of starch by amylase in the alimentary canal. It is made up of 2 molecules of glucose.

$$Glucose + Glucose \rightarrow Maltose$$

15. Sucrose is otherwise called as 'table sugar' or 'cane sugar'. It is the common sugar and is widely distributed in all photosynthetic plants. It does not exist in the body but occurs in sugarcane, pineapple, sweet potato and honey. It is made up of glucose and fructose.

Source/Occurence of Carbohydrate

S. No	Carbohydrate	Source/Occurence
1	Glucose	Broccoli, shiitake mushrooms, baked sweet potatoes, cucumber slices or chopped spinach Nuts, including almonds, peanuts and cashews, Banana, Grapes, Kiwi, cherries, dates, honey
2	Fructose	Banana, Black berry, mango, grape, Fig, Pine apple, raspberries, pear, pomegranate
3	Maltose	Cooked Sweat Potato, Molasses, Malted milk shake, Chocolate like products
4	Galactose	Honey, beats, cheese, plain Greak Yoghurt, Cherry, Kiwi, Soy sauce, plums, dry figs.
5	Lactose	Candies, Egg, ice cream, cheese, milk, cream, yoghurt.
6	Starch	Cereals, bread, cookies, cakes, pizza, potato, grains, pasta, fried food, Corn, Beans.
7	Sucrose	Grapes, figs, litches, pomegranate, cherries, Banana, Pear, plum.
8	Ribose	Mushrooms, Beef and poultry,Cheddar cheese and cream cheese. Milk, Eggs, Caviar, Anchovies, herring, sardines, Yogurt.
9	Mannose	Black currants, red currants, gooseberries, cranberries, cranberry juice, tomatoes, apples, peaches, oranges and blueberries. palm kernel, the edible oil from palm trees. Green beans, cabbage, broccoli, eggplant, turnip, green coffee beans, shiitake mushrooms and kelp.

COMMON TESTS FOR CARBOHYDRATES

Molisch's test : Sugars undergo dehydration in the presence of non-oxidizing acids like hydrochloric acid and sulphuric acid to form furfural or hydroxy methyl furfural. These compounds can react with aromatic amines such as α naphthol or phenol to give intensely coloured compounds. This reaction forms the basis of a general qualitative test for sugar.

Test : Add 2.0 ml of the given solution in a test tube and add 2 drops of an ethanolic solution of naphthol (5%). Carefully run down about 1.0 ml of Conc. sulphuric acid along the sides of the tube. Formation of a bluish violet coloured ring at the junction of two liquids or development of violet colour throughout the solution shows the presence of carbohydrate.

Anthrone test : Carbohydrates form furfural with Conc. sulphuric acid and produce bluish green colour followed by the addition of anthrone.

Test : Take 2.0 ml of the anthrone reagent (0.2% in Con H_2SO_4) in a test tube and add two drops of the test solution . Mix well. If there is no colour change, boil in a water bath for 10 min. Formation of green colour indicates the presence of carbohydrates.

Fehling's Test : If the carbohydrate has reducing group, the cupric ions present in the Fehling's reagent will be reduced to cuprous ions and even copper and this will produce a rusty brown or red precipitate.

Test : To 2 ml of fehlings solution add a few drops of the given test solution and boil. Formation of red or brown precipitate shows the presence of reducing sugar.

Benedict's Test : When the reducing sugar is boiled with Benedict's reagent which is an alkaline solution of cupric sulphate, the blue coloured cupric sulphate is reduced gradually to form insoluble cuprous oxide which may be green ,yellow, orange or red in colour, depending upon the concentration of sugar in the solution.

Test : Add 3 to 4 drops of the given test solution to 2 ml of the benedicts reagent and boil the contents. Formation of red precipitate shows the presence of reducing sugar.

Barfoed's test : This test is used to distinguish monosaccharides from disaccharides as the monosaccharides immediately give cuprous oxide red precipitate on heating (for 1-2 min) with Barfoed's reagent which contains cupric acetate in acetic acid

Test : Add 1.0 ml of given test solution to 2.0 ml of the barfoeds reagent and boil exactly for 1 minute. Formation of reddish orange precipitate shows the presence of monosaccharides. Excess boiling will give faults positive results., i.e., disaccharides also give red colour on excess boiling.

Seliwanoff's test : Keto hexoses like fructose form hydroxyl methyl furfural derivatives with hydrochloric acid and form cherry red coloured compound with resorcinol present in the Seliwanoff's reagent.

Test : Add a few drops of the given test solution to 5.0 ml of this preheated seliwanoffs reagent and boil the mixture. Formation of cherry red colour in three minutes shows the presence of ketose sugar.

Bial's test : Pentose sugar forms furfural derivatives with hydrochloric acid and then reacts with orcinol to give green coloured products.

Test : Mix 5.0 ml of Bial's reagent and 2.0 ml of the given test solution in a tube and heat in a water bath. Note the time at which any colour change is observed. Formation of green colour within 10 minutes shows the presence of a pentose.

Tollens phloroglucinol test : Add 1 ml of the test solution to 0.5 ml of phloroglucinol solution. Formation of red colour indicates the present of galactose.

Mucic acid test : Add 1 ml of the test solutoin to 0.5 ml of dilute nitric acid and heat in a boiling water bath for 90 mins. and let stand for over night. Formation of crystalline white precipitate shows the presence of galactose.

Iodine test : Iodine forms blue coloured complex with 1,4 glycosidic linkages present in polysaccharides such as starch and glycogen.

Test : Add a few drops of the given test solution to 2 drops of 0.1N HCl followed by 2 drops of iodine reagent. Formation of blue colour shows the presence of starch and brown colour shows the presence of glycogen. If the given carbohydrate solution is found to be a non-reducing one, the carbohydrate must be hydrolyzed to give reducing monosaccharide units and then all the tests should be performed to identify the components after hydrolysis.

Hydrolysis of non-reducing sugars

2.5 ml of the given solution is mixed with 5 drops of concentrated sulpuric acid and boil the contents for 5 minutes. Cool the contents and neutralize with saturated barium hydroxide. Remove the precipitate of barium sulphate by filtration and carry out all the tests in the filtrate.

Confirmatory test for carbohydrates

Phenyl hydrazine test : This is an important reaction of reducing sugars(monosaccharides and disaccharides). The aldehyde or ketonic groups present in sugar reacts with phenyl hydrazine and forms yellow crystalline products called osazones. The shape and the time of formation of osazone confirms the type of carbohydrate. The reagent is prepared by mixing 2 parts of phenyl hydrazine hydrochloride and 3 parts of sodium acetate by weight. These are thoroughly mixed in a mortar.

Test : To 2.0 ml of the given solution add about 1 spatula of the reagent mixture and boil the contents in a water bath. Note down the time taken for the formation of yellow coloured crystals. Allow the tube to cool slowly and examine the crystals microscopically. Better crystals can be obtained if the tubes are allowed to cool in a water bath.

Table : **Time of formation and the shape of osazone.**

Carbohydrate	Time of formation	Shape of osazone crystals of osazone
Mannose	1-5 mins	Yellow needle shaped crystals
Fructose	2-3 mins	Yellow needle shaped crystals
Glucose	5-7 mins	Yellow needle shaped crystals
Galactose	15-20 mins	Broken glass like crystals
Lactose	45-50 mins	Badminton ball shaped crystals.
Maltose	35-45 mins	Star shaped crystals
Arabinose	8-10 mins	Chalk powder shaped crystals
Xylose	6-7 mins	Flower shaped crystals
Sucrose	3-7 mins	Needle shaped crystals form after hydrolysis

DIGESTION AND ABSORPTION

Digestion is defined as a process that involves physical and chemical breakdown of insoluble complex food material into soluble simple food materials.

$$\text{Polysaccharides} \rightarrow\rightarrow \text{Glucose}$$
$$\text{Proteins} \rightarrow\rightarrow \text{Amino acids}$$
$$\text{Lipids} \rightarrow\rightarrow \text{Fatty acids + Glycerol}$$

There are two types of digestion, they are intracellular and extracellular digestion. The digestion occurs inside of the cell is called intracellular digestion. Eg. Phagocytosis.

Digestion taking place out side of the cell and inside of the bowl are called extracellular digestion. Eg. The digestion taking place inside of the stomach. It is also called inter cellular digestion as it occur in between the cells in the lumen of stomach.

Digestive system of Man

It includes alimentary canal and digestive glands. Six major functions take place in the digestive system are Ingestion, Secretion, Mixing and movement, Digestion, Absorption and Excretion. Alimentary canal includes mouth, tongue, pharynx, oesophagus, stomach, intestine, rectum and anus.

Mouth : It is a first part of alimentary canal. It is also called oral cavity. The tongue, teeth, and salivary glands are the accessory organs found in mouth, which aids in the digestion of food. Teeth chop food into small pieces, which are moistened by saliva before the tongue and other muscles push the food into the pharynx.

Teeth 32 small, hard organs found along the anterior and lateral edges of the mouth. Each tooth is made of a bone-like substance called dentin and covered in a layer of enamel—the hardest substance in the body.

The tongue is located on the inferior portion of the mouth just posterior and medial to the teeth. It is a small organ made up of several pairs of muscles covered in a thin, bumpy, skin-like layer. The outside of the tongue contains many rough papillae for gripping food as it is moved by the tongue's muscles. The taste buds on the surface of the tongue detect taste molecules in food and connect to nerves in the tongue to send taste information to the brain. The tongue also helps to push food toward the posterior part of the mouth for swallowing.

Surrounding the mouth are 3 sets of **salivary glands**. The salivary glands are accessory organs that produce a watery secretion known as saliva. Saliva helps to moisten food and begins the digestion of carbohydrates. The body also uses saliva to lubricate food as it passes through the mouth, pharynx, and esophagus.

Pharynx or throat is a funnel-shaped tube connected to the posterior end of the mouth. The pharynx is responsible for the passing of masses of chewed food from the mouth to the esophagus. Pharynx serves two different functions, it contains a flap of tissue known as the epiglottis that acts as a switch to route food to the esophagus and air to the larynx. **Esophagus** is a muscular tube connecting the pharynx to the stomach that is part of the upper gastrointestinal tract. It carries swallowed masses of chewed food along its length. At the inferior end of the esophagus is a muscular ring called the lower esophageal sphincter or cardiac sphincter. The function of this sphincter is to close of the end of the esophagus and trap food in the stomach.

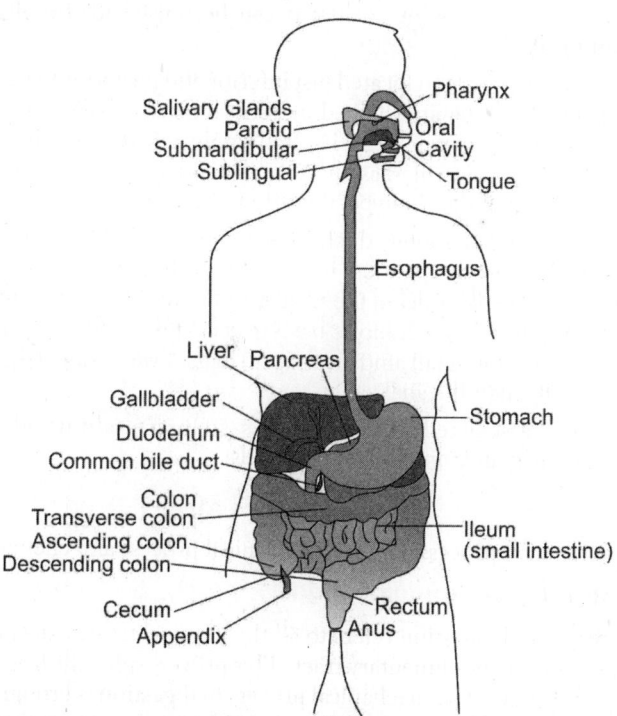

*Fig. 28 : **Anatomy of Digestive System.***

Stomach is a muscular sac that is located on the left side of the abdominal cavity, just inferior to the diaphragm. In an average person, the stomach is about the size of their two fists placed next to each other. This major organ acts as a storage tank for food so that the body has time to digest large meals properly. The stomach also contains hydrochloric acid and digestive enzymes that continue the digestion of food that began in the mouth. **Small Intestine** is a long, thin tube about 1 inch in diameter and about 10 feet long that is part of the lower gastrointestinal tract. It is located just inferior to the stomach and takes up most of the space in the abdominal cavity. The entire small intestine is coiled like a hose and the inside surface is full of many ridges and folds. These folds are used to maximize the digestion of food and absorption of nutrients. By the time food leaves the small intestine, around 90% of all nutrients have been extracted from the food that entered it. The **liver** is a roughly triangular accessory organ of the digestive system located to the right of the stomach, just inferior to the diaphragm and superior to the small intestine. The liver weighs about 3 pounds and is the second largest organ in the body. The liver has many different functions in the body, but the main function of the liver in digestion is the production of bile and its secretion into the small intestine. The **gall bladder** is a small, pear-shaped organ located just posterior to the liver. The gallbladder is used to store and recycle excess bile from the small intestine so that it can be reused for the digestion of subsequent meals.

Pancreas is a large gland located just inferior and posterior to the stomach. It is about 6 inches long and shaped like short, lumpy snake with its "head" connected to the duodenum and its "tail" pointing to the left wall of the abdominal cavity. The pancreas secretes digestive enzymes into the small intestine to complete the chemical digestion of foods.

Large Intestine is a long, thick tube about $2^1/_2$ inches in diameter and about 5 feet long. It is located just inferior to the stomach and wraps around the superior and lateral border of the small intestine. The large intestine absorbs water and contains many symbiotic bacteria that aid in the breaking down of wastes to extract some small amounts of nutrients. Feces in the large intestine exit the body through the anal canal.

The glands associated with digestive system are salivary glands, liver, pancreas, gastric glands and intestinal glands.

DIGESTION IN MAN

It involves two process, namely mechanical process and chemical process.

Mechanical process in digestion

This process of digestion refers to all the physical treatments given to the food when is inside the alimentary tract. This process splits the larger particles into smaller particle. The mechanical process of digestion is brought about by the movements of alimentary canal. It include Mastication, Deglutition, Gastric motility, Motility of small intestine, Motility of large intestine and Defecation.

1. **Mastication** : It is also called chewing. It is the rhythmic movements of the jaws, tongue and lips. It serves to breakdown the food particles into smaller masses. It is brought about by the action of teeth. This smaller particle are collected and made into a spherical ball called Bolus. The bolus is pushed backwards for swallowing.

2. **Deglutition** : After mastication, the food is swallowed into stomach. This process is called degiltition. It is divided into 3 phases. 1. Oral phase; 2. Pharyngeal Phase and Oesophagus phase

3. **Gastric motility** : The movement of the food in stomach is called gastric motility. It is of two phases. They are Hunger contraction and Digestive peristalsis.

The contraction exhibited by the empty stomach is called hunger contraction. Hunger contraction is replaced by digestive peristalsis. It occurs when the stomach contains food.

4. **Motility of Small intestine** : It is due to the rhythmicity of the circular muscles of the intestinal wall. It is characterized by alternate contractions and relaxations. It is of two types. They are Segmenting contractions and Intestinal peristalsis.

Segmenting contraction is also called pendulum movements. It is a rhythmic contraction in which a section of the intestine is divided into short segments by rings of contractions occurring at regular intervals. The rhythmic contractions are myogenic. These contractions agitate the intestinal contents. Such agitation brings about the following functions.

(i) It breaks the food materials into smaller particles.

(ii) It mixes the food with the intestinal secretions.

(iii) It constantly changes the layers of fluid in contact with the intestinal mucosa, this falicitates absorption.

Intestinal peristalsis is the contractions which travel in an aboral direction in the intestinal wall are called intestinal peristalsis. These movements are in the form of waves. Motility of villi is exhibited in two ways. They are lashing movements and rhythmical shortening and lengthening. These movements accelerate the flow of blood and lymph and that they increase absorption.

5. **Motility of large intestine** : There are two types of motility in the colon. They are the movements concerned with mixing and absorption and the movements concerned with propulsion.

Mixing and absorbing movements : These movements help to mix and absorb the contents to colon.

Propulsive movements : These movements propel the contents of colon towards the anus. These are of two types, namely colonic peristalsis and mass movements.

6. **Defaecation** : It is the explusion of faeces. The mass movements drive the faecal matter intothe colon. Generally, the rectum remains empty. The faecal matter is stored into the sigmoid and pelvic colon and not in rectum.

The faeces is the digestive waste. It has a characteristic yellow colour. This is due to bile pigments. The bad odour is due to indole, skatole, H_2S etc. An adult produces 70 to 170 gms of faecal daily. It has a pH of 7.0 to 7.5. About 25-35% of faeces is made upof solid and remaining is water.

Chemical process of digestion

It refers to the treatment of food materials with enzymes. Based on the region of digestion, it is classified into three types. They are buccal digestion or salivary digestion, Gastric digestion and Intestinal digestion

Buccal digestion : It occurs inside of the buccal cavity. Salivery enzymes play a major role. Hence it is called salivary digestion. In the mouth, food is masticated. During mastication, the food is crushed into smaller particles and mixed with secretions of the salivary gland.

Salivary glands : It is present in the buccal cavity. Man contains 3 pairs of salivary gland. They are parotid, submandibular and sublingual. Parotid glands are located just below and infront of ears. They open into the mouth through duct of stensen. Submandibular glands are present in the angles of the lower jaw. They open into the mouth by whartons duct. Sublingual glands are located below the tongue. They open into the mouth through several ducts.

The salivary glands are racemose and made upof many sac like alveoli. They unite to form large ducts which open into main ducts.

Saliva : The secretions of salivary gland are called saliva. It is a colourless and cloudy fluid. A man secretes about 1 to 1.5 L of saliva per day. Saliva is slightly acidic (pH6.02 to 7.05). Water content is 99.5%. Solid content is 0.5%. Cellular contents of saliva are yeast, bacteria, protozoa and leucocytes. Sodium chloride, potassium chloride, calcium carbonate, calcium phosphate, disodium phosphate and potassium thiocyanate are the inorganic salts of saliva. Ptyalin (Amylase), lipase, Carbonic anhydrase, lysozyme are the enzymes found in saliva.

Functions of saliva : It keeps the mouth moist and helps speech. It moistens the food and helps mastication and deglutition. It lubricates the mouth cavity and avoid thirsty. It cools down hot substances. It is essential for the appreciation of taste. Saliva washes down the food debris and these by bacteria do not grow. Saliva excretes toxic heavy metals, thiocyanate etc.,

Gastric digestion : Digestion in stomach is called stomach or gastric digestion. After swallowing, the food reaches the stomach. Here the food is treated mechanically by gastric movements as well as chemically. The chemical changes are due to gastric juices. Gastric juices are the secretion of gastric glands. There are about 35,000,000 gastric glands in man. The following gastric glands are available in man. They are

1. Mucous neck cells – Mucous

2. Chief cells or Zymogenic cells of peptic cells – Pepsin, rennin, gelatinase.

3. Oxyntic or parietal cells – hydrochloric acid.

4. Argentaffin cells – serotonin.

Gastric juice is a transparent yellow fluid. A healthy man secretes 2000 to 3000ml of gastric juice in 24 hours. It is highly acidic in nature. It has a pH of 0.5 to 1.5.

Pepsin is a proteolytic enzyme produced by peptic cells. It is secreted in an inactive form pepsinogen. It is converted to active pepsin by hydrochloric acid. It converts protein into peptone.

Rennin or rennet or chymosin is secreted by the Chief cells of gastric gland. It secretes inactive form of rennin called prorennin and activated by hydrochloric acid. It acts on casein of milk and converted to paracasein.

Prorennin →→→→ rennin

Rennin +Casein →→→→ Procasein

Procasein + Calcium →→→→ Calcium para caseinate.

Trypsin
Protein →→→→ Partical digestion to amino acid

Chymotrypsin
Protein →→→→→→ Small Polypeptide + Amino acid

Carboxypeptidace
Peptides →→→→→→→ Amino acid

Elastase
Elastin →→→→ Elastic Fibre

Amylase
Starch →→→→ Glucose

Lipase
Lipid →→→→ Fatty acid

Invertase/Sucrase
Sucrose →→→→→→→ Glucose + Fructose

Maltase
Maltose →→→→ Glucose + Glucose

Lactase
Lactose →→→→ Glucose + Galactose

RNase/DNase
RNA/DNA →→→→→→ Nucleotides

Intestinal digestion

Small intestine, especially duodenum, is significant, because it receives pancreatic juice from pancrease, bile from liver and intestinal juice from intestinal glands. Pancreatic juice is secreted by acini cells of pancrease. It is a colourless odourless alkaline fluid. It is isotonic with blood. The daily secretion is about one liter. The Important enzymes present in the pancreatic juice are trypsin, chymotrypsin, carboxy peptidase, pancreopeptidase or elastase, amylase, lipase, sucrose, maltase, lactase and nuclease.

Liver

It is a digestive gland. It is the largest gland in the human body. Liver contains two lobes. It is formed of hepatic cells. Numerous capillaries are present in between hepatic cells. These capillaries are called bile capillaries. Hepatic duct connects all capillaries through blind tube and bigger vessels.

Functions : It is the largest chemical factory of human body. It synthesizes glycogen from glucose. It stores glycogen. It secretes bile. It synthesizes urea. It modifies waste and toxic substances suitable for excretion through bile or urine. It helps in the absorption of fats. It produces RBC in the foetal life. It destroys the dead RBC. It removes bilirubin from blood. It manufactures plasma protein. It produces prothrombin and fibrinogen.

Gall Bladder : It is a digestive gland. It is located on the upper surface of liver. A duct called cystic duct arises from gall bladder. It opens into hepatic duct. It functions as a store house of bile.

Bile : It is a product of secretion and excretion of liver. It is produced by the parenchymal cells of the liver. It is stored in the gall bladder. It is a clear, golden yellow or greenish fluid. It is slightly viscous. It has a bitter taste and alkaline in nature. Daily secretion is from 500 ml to 1 liter.

Bile salt : It is a sodium and potassium salt of glycocholic acid and taurocholic acid. They emulsify the fat and thereby increases the surface area of the surface exposed to pancreatic juice. It activates the enzymes cholesterol esterase and pancreatic lipase. They facilitate the absorption of fat soluble vitamins. They promote the secretary power of liver cells. They activate inactive lipase. They increase intestinal motility.

Bile pigment : They are metabolic waste products of the degradation of heme. There are two types of bile pigments. They are bilirubin and biliverdin.

Intestinal juice or Succus entericus

It is secreted by two different types of glands in the intestine. They are

1. Glands of Brunner or duodenal glands

2. Intestinal glands or crypts of Liberkhnn

The following enzymes are present in the intestinal juice. They are enteropeptidase, erepsin (dipeptidases and aminopeptidases), arginase, amylase, sucrose, maltase, isomaltase, lactase, lipase, alkaline phosphatase, cholesterol esterase, lecithinase, phospholipase.

ABSORPTION

It is a process by which the end products of digestion are transported from the lumen of alimentary canal to the blood stream through the intestinal wall.

Place of absorption

'Simple substances' which donot require digestion are absorbed in stomach eg. Water, alcohol, saline, glucose. Small intestine is the main place for absorption. The small intestine has two important adaptations for effective absorption. They are

1. It has great length about 25 feet and is much coiled.

2. The intestinal mucosa is produced into finger like projections called villi into the lumen. Human intestine contains as many as 50,000, 000 villi.

Large intestine absorps water, saline, alcohol and certain drugs.

Absorption of carbohydrates

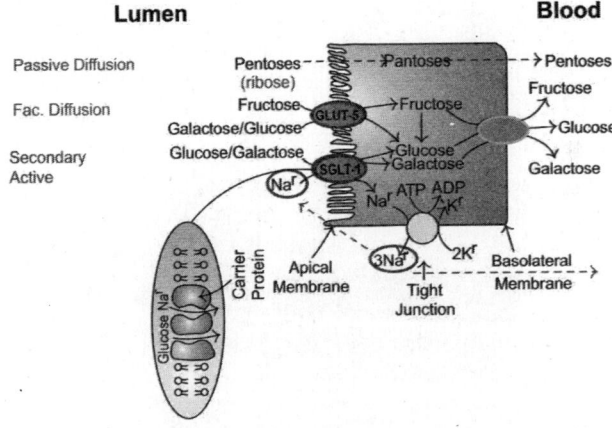

*Fig. 29 : **Absorption of Carbohydrate.***

It is the intake of monosaccharides from the lumen into the blood. Maximum rate of glucose absorption is 120g/hr. Carbohydrate is absorbed in the form of glucose, galactose, fructose, manose and pentose. Glucose and galactose are absorbed through active transport. Plasma membrane of intestinal wall has carrier protein sodium dependent glucose transporters (SGLT). This protein has two binding sites, one for glucose another one for Na^+. it binds with Na^+ and glucose & move towards cytoplasm of the cell which brings Na+ and glucose facing the cytoplasm. Here both glucose and Na^+ are released. Glucose moves along the cytoplasm and diffuses in the blood. The Na^+ is transported back to the lumen through sodium potassium pump. The simultaneous transport of Na^+ and glucose are called Co-Transport. It is transported in a same direction, it is called symport. SGLT also transports galactose. Fructose is transported by the carrier protein GLUT (Glucose transporters). It is independent of Na^+. Other monosaccharides are transported by diffusion (**Fig.29**).

Absorption of protein

Absorption of protein is the intake of amino acids from the lumen of intestine into the blood. They are transported into the portal system of blood.

Most of the protein absorption occurs in the jejuneum and some occurs in ileuem. Amino acids are transported through plasmamambrane. It is transported by active transport. Its transport is mediated by carrier protein present in the plasmamambrane of the intestinal cells. Amino acids move along the cytoplasm and diffuses into the blood. L forms of amino acids are transported actively and rapidly. Some small peptides are also enters in blood. Peptides are transported along with sodium (**Fig.30**).

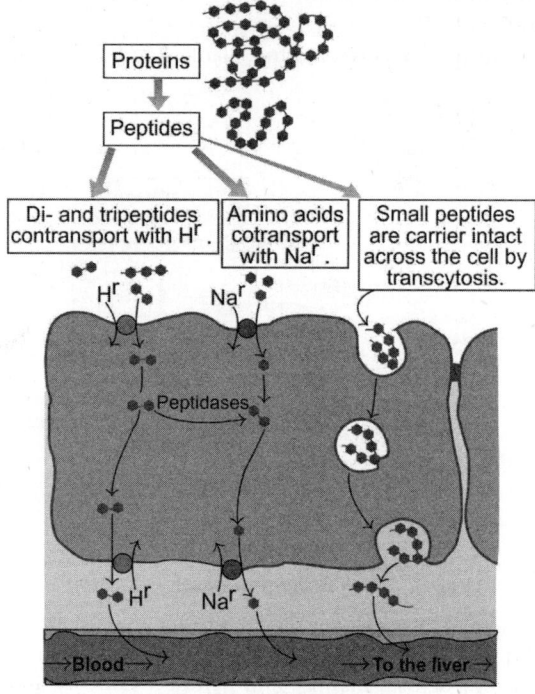

Fig. 30 : ***Absorption of Protein***

Absorption of lipid

The major products of lipid digestion are fatty acids and 2-monoglycerides. They enter the enterocyte by simple diffusion across the plasma membrane. A considerable fraction of the fatty acids also enter the enterocyte via a specific fatty acid transporter protein in the membrane. Lipids are transported from the enterocyte into blood by a mechanism distinctly different from what we've seen for monosaccharides and amino acids. Once inside the enterocyte, fatty acids and monoglyceride are transported into the endoplasmic reticulum, where they are used to synthesize triglyeride. Beginning in the endoplasmic reticulum and continuing in the Golgi, triglyceride is packaged with cholesterol, lipoproteins and other lipids into particles called chylomicrons. Chylomicrons are extruded from the Golgi into exocytotic vesicles, which are transported to the basolateral aspect of the enterocyte. The vesicles fuse with the plasma membrane and undergo exocytosis, dumping the chylomicrons into the space outside the cells. Because chylomicrons are particles, virtually all steps in this pathway can be visualized using an electron microscope, as the montage of images to the right demonstrates. Transport of lipids into the circulation is also different from what occurs with sugars and amino acids. Instead of being absorbed directly into capillary blood, chylomicrons are transported first into the lymphatic vessel that penetrates into each villus. Chylomicron-rich lymph then drains into the system lymphatic system, which rapidly flows into blood. Blood-borne chylomicrons are rapidly disassembled and their constitutent lipids utilized throughout the body.

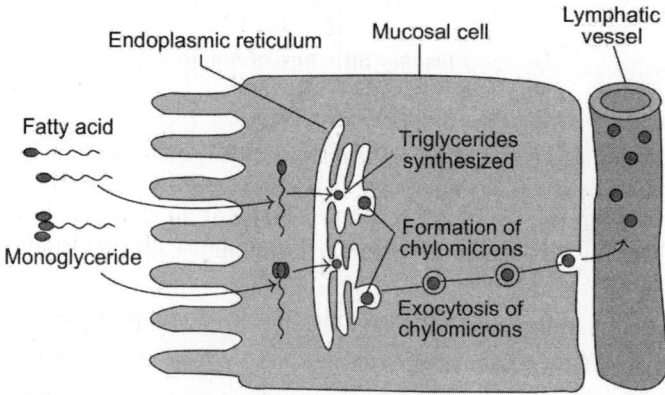

*Fig. 31 : **Absorption of lipid.***

Definition

The amino acids are the building blocks for proteins. All proteins studied are made from the twenty standard amino acids. Amino acids are the simplest units of a protein molecule and they form the building blocks of protein structure.

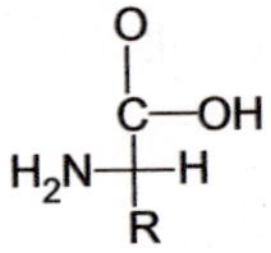

$$H_2N—\underset{\underset{R}{|}}{\overset{\overset{\displaystyle O}{|}}{\underset{|}{C}}}—H$$

Fig. 32 : Fischer Structure of Amino acid.

Structure of Amino acid

An amino acid is an amino carboxylic acid. R is the side chain or residue and it represents the group other than $-NH_2$ and $-COOH$. It may be a hydrogen atom (H) or a methyl group ($-CH_3$) or an aliphatic group or an aromatic group or a heterocyclic group. The amino acids are classified based on the nature of R groups.

Generally, amino acids have the following structural properties:

A carbon (the alpha carbon)   **A hydrogen atom (H)**

A Carboxyl group (–COOH)   **An Amino group (–NH₃)**

A "variable" group or "R" group (Fig.32)

All amino acids have the alpha carbon bonded to a hydrogen atom, carboxyl group, and amino group. The "R" group varies among amino acids and determines the differences between these protein monomers. The amino acid sequence of a protein is determined by the information found in the cellular genetic code. The genetic code is the sequence of nucleotide bases in nucleic acids (DNA and RNA) that code for amino acids. These gene codes not only determine the order of amino acids in a protein, but they also determine a protein's structure and function.

D and L amino acids

Based on the position of amino group on the asymmetric carbon atom, amino acids exist in two types. They are D and L amino acids.

L-amino acid
CO-R-N goes anticlockwise

D-amino acid
CO-R-N goes clockwise

Fig. 33 : L and D Amino acid.

The amino acid having the NH_2 group on the right is called D-amino acid. The amino acid having the NH_2 group on the left is called L-amino acid. These isomers are the mirror images of each other. All amino acids are α amino acids because the NH_2 group is attached to a carbon atom which is next to the COOH group. Examination of the structure of an amino acid except glycine, reveals that the α carbon atom has four different groups attached to it, thus making it asymmetric. Because of the presence of asymmetric carbon atom, amino acids exist in two optically active forms, dextrorotatory and levorotatory.

Dextrorotatory, compounds rotate plane polarised light in the clockwise direction. Levorotatory, compounds rotate plane polarised light in the anti clockwise direction. The direction of optical rotation of an amino acids indicated by the symbol + and - (+ indicates dextro and - indicate levo).

It has been found that L-amino acids are more common than D forms and most of the naturally occuring amino acids are L-amino acids. Therefore L-amino acids are called natural amino acids. Since the L amino acids are more common, the letter "L" is usually omitted, while representing L-amino acids.Amino acids are widely distributed in plants and animals.

PROPERTIES OF AMINO ACIDS

Physical

Amino acids are white crystalline substances.

Most of them are soluble in water and insoluble in non-polar organic solvents (e.g., chloroform and ether).

Aliphatic and aromatic amino acids particularly those having several carbon atoms have limited solubility in water but readily soluble in polar organic solvents.

They have high melting points varying from 200-300°C or even more.

They are tasteless, sweet or bitter. Some are having good flavour.

Amphoteric nature of amino acids as they contain both acidic (COOH) and basic (NH_2) groups. The amino acids possessing both positive and negative charges are called zwitterions.

Isomerism : All amino acids except proline, found in protein are α-amino acids because NH_2 group is attached to the α-carbon atom, which is next to the COOH group. Examination of the structure of amino acids reveals that except glycine, all other amino acids possess asymmetric carbon atom at the position.

Chemical Properties

Reaction with formaldehyde (Formal titration) : An amino acid solution is treated with excess of neutralized formaldehyde solution; the amino group combines with formaldehyde forming dimethylol amino acid which is an amino acid formaldehyde complex. Hence the amino group is protected and the proton released is titrated against alkali. This method is used to find out the amount of total free amino acids in plant samples.

Reaction with nitrous acid : Nitrous acid reacts with the amino group of amino acids to form the corresponding hydroxyacids and liberate nitrogen gas.

Reaction with ninhydrin : Ninhydrin is a strong oxidizing agent. When a solution of amino acid is boiled with ninhydrin, the amino acid is oxidatively deaminated to produce ammonia and a ketoacid. The keto acid is decarboxylated to produce an aldehyde with one carbon atom less than the parent amino acid. The net reaction is that ninhydrin oxidatively deaminates and decarboxylates α-amino acids to CO_2, NH_3 and an aldehyde. The reduced ninhydrin then reacts with the liberated ammonia and another molecule of intact ninhydrin to produce a purple coloured compound known as Ruhemann's purple.

- This ninhydrin reaction is employed in the quantitative determination of amino acids. Proteins and peptides that have free amino group(s) (in the side chain) will also react and give colour with ninhydrin.

Decarboxylation

The carboxyl group of amino acids is decarboxylated to yield the corresponding amines.

CLASSIFICATION OF AMINO ACIDS BASED ON POLARITY

The protein amino acids are classified according to the chemical nature of their R groups as aliphatic, aromatic, heterocyclic and sulphur containing amino acids. More meaningful classification of amino acids is based on the polarity of the R groups. The polarity of the R groups varies widely from totally non-polar to highly polar. The 20 amino acids are classified into four main classes.

Amino acids with non-polar or hydrophobic, aliphatic R groups

This group of amino acids includes glycine, alanine, valine, leucine. Isoleucine and proline. The hydrocarbon R groups are non-polar and hydrophobic. The side chains of alanine, valine, leucine and isoleucine are important in promoting hydrophobic interactions within protein structures. On the other hand, the imino group of proline is held in a rigid conformation and reduces the structural flexibility of the protein.

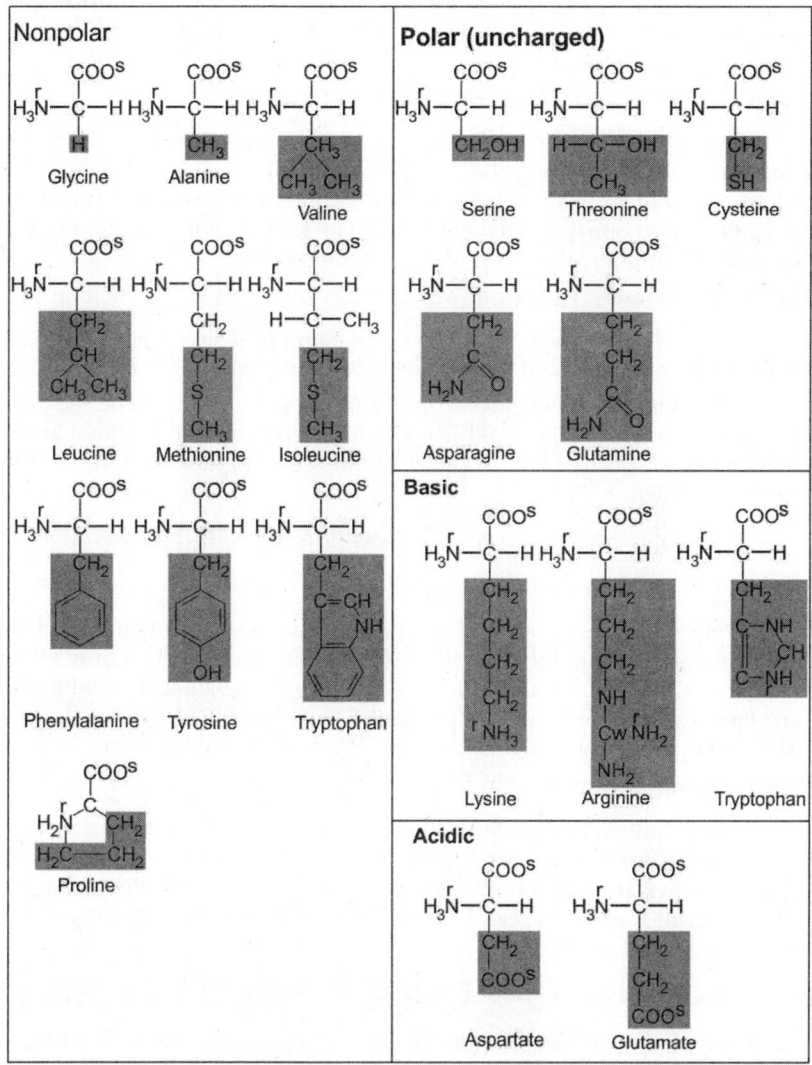

*Fig. 34 : **Structure of Amino acid.***

Amino acids with non-polar aromatic R groups

This group includes phenylalanine, tyrosine and tryptophan. All these amino acids participate in hydrophobic interactions, which is stronger than aliphatic R groups because of stacking one another. Tyrosine and tryptophan are more polar than phenylalanine due to the presence of hydroxyl group in tyrosine and nitrogen in the indole ring of tryptophan. The absorption of ultraviolet (UV) light at 280 nm by tyrosine, tryptophan and to a lesser extent by phenylalanine is responsible for the characteristic strong absorbance of light by proteins. This property is exploited in the characterization and quantification of proteins.

Amino acids with polar, uncharged R groups

This group of amino acids includes serine, threonine, cysteine, methionine, asparagine and glutamine. The hydroxyl group of serine and threonine, the sulphur atom of cysteine and methionine and the amide group of asparagine and glutamine, contribute to the polarity. The R groups of these amino acids are more hydrophilic than the non-polar amino acids.

Amino acids with charged R groups

Acidic : The two amino acids with acidic R groups are aspartic and glutamic acids. These amino acids have a net negative charge at pH 7.0.

Basic : This group includes lysine, arginine and histidine. The R groups have a net positive charge at pH 7.0. The lysine has a second E-amino group; arginine has a positively charged guanidino group; and histidine has an imidazole group.

NUTRITIONAL CLASSIFICATION OF AMINO ACIDS

Essential amino acids

Certain amino acids can not be synthesized by the living organisms. They must be compulsarily included in the diet for normal health.These amino acids are called essential amino acids. For human being about 10 amino acids are considered as essential.eg, Histidine, Isoleucine, Leucine, Lysine, Methionine, Phenylalanine, Threonine, Tryptophan, Valine, Alanine.

Non-essential amino acids

Certain amino acids can be synthesized in the cells from essential amino acids or from other compounds. So these amino acids need not be included in the diet. They are called non-essential amino acids. Eg. Arginine, Aspartic acid, Cysteine, Glutamin acid, Glycine, Proline, Serine, Tyrosine, Aspargine, Selenocysteine, Pyrrolysine.

CLASSIFICATION BASED ON METABOLIC RATE

The carbon skeleton of amino acid serve as a precursor for the synthesis of glucose or fats or both. Based on this amino acids are classified as glycogenic amino acids, ketogenic amino acid and glycogenic and ketogenic amino acids.

Glycogenic amino acids – these amino acids form glucose or glycogen. Eg. Alanine, aspartate etc.,

Ketogenic amino acids – these amino acids form fat. Eg. Leucine and Lysine.

Glycogenic and ketogenic amino acids – Some amino acids serve as a precursor for both glucose and fat. Eg. Phenylalanine, tryptophan, Tyrosine and Isoleusine.

PROTEINOGENIC AMINO ACID

Proteinogenic amino acids are amino acids that are incorporated biosynthetically into proteins during translation. The word *"proteinogenic"* means *"protein creating"*. Throughout known life, there are 22 genetically encoded (proteinogenic) amino acids, 20 in the standard genetic code and an additional 2 that can be incorporated by special translation mechanism.

Selenocysteine (Sec) and pyrrolysine (Pyl) are rare amino acids that are cotranslationally inserted into proteins and known as the 21st and 22nd amino acids in the genetic code. Sec and Pyl are encoded by UGA and UAG codons, respectively, which normally serve as stop signals.

Functions of Non protein amino acids

Certain amino acids which do not exist in proteins are called non protein amino acids eg. Ornithine and β-alanine etc.

- In humans, non-protein amino acids also have important roles as metabolic intermediates, such as in the biosynthesis of the neurotransmitter gamma-amino-butyric acid (GABA). Many amino acids are used to synthesize other molecules.

- Tryptophan is a precursor of the neurotransmitter serotonin.

- Tyrosine (and its precursor phenylalanine) are precursors of the catecholamine neurotransmitters dopamine, epinephrine and norepinephrine and various trace amines.

- Phenylalanine is a precursor of phenethylamine and tyrosine in humans. In plants, it is a precursor of various phenylpropanoids, which are important in plant metabolism.

- Glycine is a precursor of porphyrins such as heme.

- Arginine is a precursor of nitric oxide.

- Ornithine and S-adenosylmethionine are precursors of polyamines.

- Aspartate, glycine, and glutamine are precursors of nucleotides.

CHAPTER 3

PROTEINS

Definition

Proteins can be defined as the high molecular weight polymers composed of α-amino acids united to one another by peptide linkage (–CO–NH–). Proteins are the major consitituents of all living organisms. They contain carbon, hydrogen, nitrogen, oxygen and sulphur. The term protein is derived from Greek word Proteios, which means primary or holding first place. Protein is the essential constituent of living cells. Protein make upto 12% of the protoplasm. They are not only responsible for comprising the structure of the cell but are concerned with every function of the cell including those of respiration, catalysis of reactions by enzymes, transport of materials, regulation of metabolism, and defense actions. The foods rich in proteins are known as **'body building foods'**.

CLASSIFICATION

Classification based on source of availability

The proteins have been traditionally divided into two well-defined groups. They are animal proteins and plant proteins. **Animal proteins** are the proteins derived from animal sources such as eggs, milk, meat and fish. They are usually called higher-quality proteins because they contain adequate amounts of all the essential amino acids. **Plant proteins** are derived from plants. They are called lower-quality proteins since they have a low content of one or more of the essential amino acids. Methionine, lysine, threonine and tryptophan are available in low quantities.

Classification based on shape

1. **Globular or Corpuscular Proteins :** These are spherical or ovoid shape. These are usually soluble in water or in aqueous media containing acids, bases, salts or alcohol and diffuse readily. Globular proteins are more complex in conformation than fibrous proteins. They have a far greater variety of biological functions. Tertiary and quaternary structures are usually associated with this class of proteins. Protein hormones, blood transport proteins, antibodies and nutrient storage proteins are the examples for globular proteins.

2. **Fibrous or Fibrillar Proteins :** These proteins resemble long ribbons or fibres in shape. These are mainly of animal origin. They are insoluble in all common solvents such as water, dilute acids, alkalies and salts and also in organic solvents. Most fibrous proteins serve in a structural or protective role. The fibrous proteins are extremely strong and possess two important properties which are characteristic of the elastomers. These are:

(a) They can stretch and later recoil to their original length.

(b) They have a tendency to creep, i.e., if stretched for a long time, their basic length increases and equals the stretched length but, if the tension on the two ends of the fibril is relaxed, they creep to their shorter and shorter length. It is a heterogeneous group and includes the proteins of connective tissues, bones, blood vessels, skin, hair, nails, horns, hoofs, wool and silk. The important examples are:

(i) Collagens : These are of mesenchymal origin and form the major proteins of white connective tissues and of bone.

(ii) Elastins : They are of mesenchymal origin; form the major constituents of yellow elastic tissues (ligaments, blood vessels); differ from collagens in not being converted to soluble gelatins.

(iii) Keratins : These are of ectodermal origin; form the major constituents of epithelial tissues (skin, hair, feathers, horns, hoofs, nails); usually contain large amounts of sulfur in the form of cystine– human hair has about 14% cystine.

(iv) Fibroin : It is the principal constituent of the fibres of silk; composed mainly of glycine, alanine and serine units.

Classification of protein based on composition and solubility

Based on composition and solubility proteins are classified as follows .

I. Simple protein

These proteins on hydrolysis yield only L-amino acids. (eg). albumin, globulin. These are further classified mainly on their solubility basis as follows:

1. Protamines and histones : These are basic proteins. It occurs in animals, mainly in sperm cells. They possess simplest structure and lowest molecular weight (approximately 5,000). They are soluble in water. They are not coagulated by heat. They are strongly basic in character. They form salts with mineral acids and nucleic proteins. Protamines are virtually devoid of sulfur and aromatic amino acids. Histones are weaker bases, therefore, insoluble in NH_4OH solution. Protamines are soluble in NH_4OH. E.g., protamines—salmon sperm; histones—nucleohistones of nuclei.

2. Albumins : These are widely distributed in nature. They are more abundant in seeds. They are soluble in water and dilute solutions of acids, bases and salts. They are precipitated with a saturated solution of an acid salt like $(NH_4)_2SO_4$. They are coagulated by heat. E.g., ovalbumin from white of egg, serum albumin from blood plasma.

3. Globulins : These are of two types of globulins. They are pseudoglobulins and euglobulins. Euglobulins are more widely distributed in nature than the pseudoglobulins. They are either soluble (pseudoglobulins) or insoluble (euglobulins) in water. They are precipitated with half saturated solution of $(NH_4)_2SO_4$. They are coagulated by heat. e.g., pseudoglobulins—pseudoglobulin of milk whey.Eg., Euglobulins— serum globulin from blood plasma, ovoglobulin from eggwhite.

4. Glutelins : These have been isolated only from plant seeds. They are insoluble in water, dilute salt solutions and alcohol solutions. They are soluble in dilute acids and alkalies. They are coagulated by heat. E.g., glutenin from wheat, oryzenin from rice, etc.

5. **Prolamines P:** These have also been isolated only from plant seeds. They are insoluble in water and dilute salt solutions. They are soluble in dilute acids and alkalies and also in 60–80% alcohol solutions. They are not coagulated by heat E.g., gliadin from wheat, hordein from oat, etc.

6. **Scleroproteins or Albuminoids :** These occur almost entirely in animals. They are commonly known as the 'animal skeleton proteins'. They are insoluble in water, dilute solution of acids, bases and salts and also in 60–80% alcohol solutions. They are not attacked by enzymes. E.g., Collagen of bones, elastin in ligaments, keratin in hair and horry tissues and fibroin of silk.

II. Conjugated protein

These are proteins composed of simple proteins combined with non-protein part called as prosthetic groups. They are further subdivided into:

1. **Nucleo protein :** Proteins present along with nucleic acids. (eg) Histones and Protamines.

2. **Phosphoprotein :** These are protein containing phosphoric acid (eg) casein of milk.

3. **Glycoprotein :** These are proteins containing carbohydrate moiety as prosthetic group. (eg.) Gonadotropic hormone, mucous glycoprotein mucin (saliva).

4. **Chromoprotein :** These proteins contain heterocyclic compounds like porphyrins as the prosthetic group. (eg) Haemoglobin and Myoglobin.

5. **Lipoproteins :** These are proteins conjugated with lipids (Eg) Chlyomicron, very low density lipoprotein (VLDL), low density lipoprotein (LDL) and high density lipoprotein (HDL).

6. **Metalloproteins :** These proteins contain metal as prosthetic group (Eg) Siderophilin (Fe) and Ceruloplasmin (Cu).

III. Derived proteins

These are proteins derived from the simple and conjugated proteins by the action of acids, alkalies or enzymes.

They are the products resulting from partial to complete hydrolysis of proteins. (eg.) proteoses, peptones and peptides.

1. **Primary derived proteins :** These are derivatives of proteins in which the size of protein molecule is not altered materially.

    (a) **Proteans :** Insoluble in water; appear as first product produced by the action of acids, enzymes or water on proteins. e.g., edestan derived from edestin and myosan derived from myosin.

    (b) **Metaproteins or Infraproteins :** Insoluble in water but soluble in dilute acids or alkalies. They are produced by further action of acid or alkali on proteins at about 30–60°C. e.g., acid and alkali metaproteins.

(c) Coagulated Proteins : Insoluble in water; produced by the action of heat or alcohol on proteins. e.g., coagulated eggwhite.

2. Secondary derived proteins : These are derivatives of proteins in which the hydrolysis has certainly occurred. The molecules are, as a rule, smaller than the original proteins.

CLASSIFICATION BASED ON FUNCTION

Proteins are the fundamental constituents of all protoplasm and are involved in the structure and functions of living cells. Proteins have many different biological functions.

1. Catalytic proteins : Enzymes are proteins which have catalytic power. They enhance the rate of biochemical reactions. (eg.) Amylase, Protease etc.

2. Nucleoproteins : Hi stones are basic proteins found in association with nucleic acids. They serve as carriers of genetic characters and hence govern inheritance of characters.

3. Hormonal proteins : Some hormones are proteins which regulate numerous physiological functions. (eg) growth hormone, insulin and glucagon.

4. Storage proteins : They have the function of storing amino acids as nutrients and as building blocks for the growing embryo. (eg) Casein of milk, ovalbumin of egg white.

5. Transport proteins : They are capable of binding and transporting specific types of molecules via blood. (eg) haemoglobin and albumin.

6. Contractile proteins : Proteins such as actin and myosin in skeletal muscle function as essential elements in contractile and motile systems.

7. Defensive proteins : Some proteins have protective or defensive functions. The blood proteins – thrombin and fibrinogen participate in blood clotting. Antibodies or immunoglobulins are protective proteins which prevent the onset of diseases in the body.

8. Structural proteins : Proteins such as collagen, keratin etc. serve as structural elements.

9. Toxic proteins : Ricin of caster bean, diphtheria toxin and botulinum toxin represents another group of proteins which cause dysfunctions and disorders in the body.

STRUCTURE OF PROTEIN

Proteins are made up of amino acids. They are the polymers of amino acids. Amino acids are the building blocks of proteins.

Protein is made up of one or more polypeptide chains

Many proteins, such as myoglobin, consist of a single polypeptide chain. Others contain two or more chains, which may be either identical or different. For example haemoglobin is formed of 4 polypeptide chains, of which two α chains are of one kind and the other two β chains are of another kind.

Peptide bonds

In proteins, amino acids are linked together by linkages called peptide bonds. The carboxyl group of one amino acid is joined to α amino group of another amino acid by a peptide bond. The peptide bond is also called as the amide bond. The two amino acids, joined by a peptide bond, constitute a dipeptide. The dipeptide is formed by simple condensation reaction. The product formed by a peptide bond is called a peptide. The compound formed by the linking of three amino acids is called as tripeptide. A peptide formed of less than 10 amino acids constitute an oligopeptide. More than 10 amino acids join together to form a polypeptide chain.

N and C terminal ends of protein

An amino acid in a polypeptide is called a residue. A polypeptide have two ends, namely amino and carboxyl terminal end. The end of the polypeptide chain containing amino group is called amino terminal or N-terminal. The end of the polypeptide chain containing carboxyl group is called carboxyl terminal or C-terminal. The terminal amino acid with the free amino group is called N-terminal amino acid and the terminal amino acid with the free carboyl group is called C-terminal amino acid.

Sequence of amino acid

The sequence of amino acid varied from one protein to other. The amino acid sequence in polypeptides with 30-40 amino acids can be determined by **Edman reaction**. For polypeptides containing more than 40 amino acids, both enzymatic and chemical methods are employed to break polypeptide chains into smaller peptides. The enzyme, trypsin hydrolyses the peptide bond on the carboxyl side of the basic amino acid residues of lysine or arginine. The chemical reagent, cyanogen bromide cleaves peptide bond on the carboxyl side of methionine residues. The hydrolyzed peptides are separated and the amino acid sequence is determined by Edman reaction. The hydrolysis of the original polypeptide by two different methods separately gives overlapping regions, from which the sequence is derived.

Chemical bonds involved in protein structure

Refer – Tertiary structure of protein – Page No. 47.

PROPERTIES OF PROTEINS

Physical properties

1. Colour and taste : Proteins are colourless and usually tasteless. These are homogeneous and crystalline.

2. Solubility : Solubility of proteins is influenced by pH. Solubility is lowest at isoelectric point and increased with increasing acidity of alkalinity.

3. Optical activity : All protein solutions rotate the plane polarised light to the left i.e. these are levorotatory.

4. **Colloidal nature :** Because of their giant size, the proteins exhibit many colloidal properties are : **(i)** Their diffusion rate is extermely low. **(ii)** They may produce considerable light-scattering in solution, thus resulting in visible turbidity (Tyndall effect).

5. The comparatively week forces responsible for maintaining secondary, tertiary and quarternary structure of proteins are readily distrupted with resulting loss of biologic activity. This distruption of native structure is termed denaturation. Physically, denaturation may be viewed as randomizing the conformation of a polypeptide chain without affecting its primary structure

Denaturation of protein

The comparatively weak forces responsible for maintaining secondary, tertiary and quaternary structure of proteins are readily disrupted with resulting loss of biological activity. This disruption of native structure is termed denaturation.

Physical and chemical factors are involved in the denaturation of protein:

- Heat and UV radiation supply kinetic energy to protein molecules causing their atoms to vibrate rapidly, thus disrupting the relatively weak hydrogen bonds and salt linkages. This results in denaturation of protein leading to coagulation. Enzymes easily digest denatured or coagulated proteins.

- Organic solvents such as ethyl alcohol and acetone are capable of forming intermolecular hydrogen bonds with protein disrupting the intramolecular hydrogen bonding. This causes precipitation of protein.

- Acidic and basic reagents cause changes in pH, which alter the charges present on the side chain of protein disrupting the salt linkages.

- Salts of heavy metal ions (Hg^{2+}, Pb^{2+}) form very strong bonds with carboxylate anions of aspartate and glutamate thus disturbing the salt linkages. This property makes some of the heavy metal salts suitable for use as antiseptics.

Renaturation

Renaturation refers to the attainment of an original, regular three dimensional functional protein after its denaturation. When active pancreatic ribonuclease A is treated with 8M urea or mercaptoethanol, it is converted to an inactive, denatured molecule. When urea or mercaptoethanol is removed, it attains its native (active) conformation.

Chemical propertiesHydrolysis

1. Hydrolysis

(a) By acidic agents : **Proteins upon hydrolysis with concentrated mineral acids such as, HCl yield amino acids in the form of their hydrochlorides.**

(b) By proteolytic enzymes : Under relatively mild conditions of temperature and acidity, certain proteolytic enzymes like **pepsin** and **trypsin** hydrolyse the proteins. Enzyme hydrolysis is used for the isolation of certain amino acids like tryptophan. Two important drawbacks with this type of hydrolysis are: (i) It requires prolonged incubation and (ii) Hydrolysis may be incomplete.

Colour Reactions of Proteins

The colour reactions of proteins are of importance in the qualitative detection and quantitative estimation of proteins and their constituent amino acids. Biuret test is extensively used as a test to detect proteins in biological materials.

Biuret Reaction : A compound, which is having more than one peptide bond when treated with Biuret reagent, produces a violet colour. This is due to the formation of coordination complex between four nitrogen atoms of two polypeptide chains and one copper atom.

Xanthoproteic Reaction : Addition of concentrated nitric acid to protein produces yellow colour on heating, the colour changes to orange when the solution is made alkaline. The yellow stains upon the skin caused by nitric acid are the result of this xanthoproteic reaction. This is due to the nitration of the phenyl rings of aromatic amino acids.

Hopkins-Cole Reaction : Indole ring of tryptophan reacts with glacial acetic acid in the presence of concentrated sulphuric acid and forms a purple coloured product. Glacial acetic acid reacts with concentrated sulphuric acid and forms glyoxalic acid, which in turn reacts with indole ring of tryptophan in the presence of sulphuric acid forming a purple coloured product.

Protein structure / Structural organization of Protein

The architecture of protein molecule is complex but well organised. To understand this, a clear idea of certain basic details regarding the mode of arrangement of the structural units inside the molecule is necessary. Linder strom - Lang suggest four types of structural organisation for proteins. They are

1. Primary structure   2. Secondary structure
3. Tertiary Structure and   4. Quarternary strucutre.

Primary structure

Primary structure is ultimately responsible for the native structure of the protein. Frederick Sanger in 1953 determined the complete amino acid sequence of insulin for the first time. The primary structure of protein is defined as the sequence of amino acid residues making up its polypeptide chain. The protein may be formed of one or more polypeptide chains. The amino acids are arranged in specific sequence in these polypeptide chain. The amino acid residues are linked by peptide bonds. The peptide bond is formed between the carboxyl group of one amino acid and the amino group of adjacent amino acid. Some times the adjacent polypeptide chains are linked by disulphide bonds. Each polypeptide chain of any length has at one end a N-terminal amino acid containing free amino group and at the other end a C-terminal amino acid containing a free carboxyl group. The amino acids in a polypeptide chain are numbered from the N-terminal end.

The primary structure has the following salient features

1. Primary structure refers to the linear sequence of amino acid residues.

2. The proteins are linear and unfolded

3. The protein is formed of one or more polypeptide chains.

4. The amino acid residues are linked by repeating polypeptide bonds.

5. The adjacent polypeptide chains are linked by disulphide bonds.

6. Most of the structural proteins which are in the form of fibres exhibit primary structure.

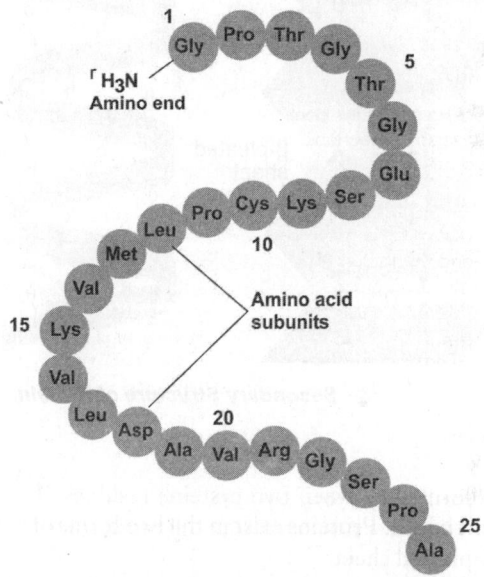

*Fig. 35 : **Primary Structure of Protein.***

7. The primary structure provides information on the number and proportion of different amino acids in a protein. Primary structures of a large number of proteins have been determined. eg. (i) human insulin has 51 amino acids distributed in two poly peptide chains. A chain-31 amino acids, B chain-20 amino acids and the polypeptides are linked by disulphide bridges. (ii) cytochrome C contains 104 amino acids. (iii) human serum albumin contains 584 amino acids.

Secondary structure

The conformation of the peptide chain by way of folding or coiling consisting of a helically coiled is called secondary structure. It is a shape taken up by the polypeptide chain. zig-zag linear or mixed form. It results from the steric relationship between amino acids located relatively near to each other in the peptide chain. The linkages or bonds involved in the secondary structure formation are hydrogen bonds and disulphide bonds.

I. Hydrogen bond

These are weak, low energy non-covalent bonds sharing single hydrogen by two electronegative atoms such as O and N. Hydrogen bonds are formed in secondary structure by sharing H-atoms between oxygen of one peptide and nitrogen of of different peptide bonds. The hydrogen bonds in secondary structure may form either a α-helix or β-pleated sheet structure.

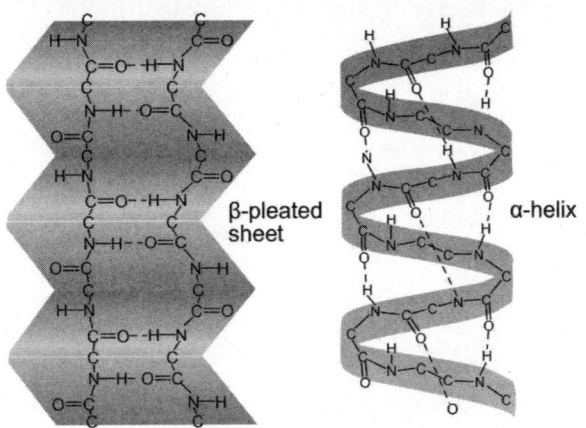

Fig. 36 : **Secondary Structure of Protein.**

II. Disulphide bond

These are formed between two cysteine residues. They are strong, high energy covalent bonds. Proteins exist in the two forms of secondary structure, α-helix and β-pleated sheet.

α-Helix

It was proposed by Pauling and Corey in 1951. A polypeptide chain forms regular helical coils called α-helix. These coils are stabilized by hydrogen bonds between carbonyl oxygen of first amino and amide N of fourth amino acid residues. Thus in α-helix intra chain hydrogen bonding is present. The α-helices can be either right handed or left handed. Left handed α-helix is less stable because of the steric interference between the carbonyl group and the side chains. Only the right handed α-helix has been found in protein structure. Each turn of the helix contains 3.6 amino acid residues. The distance between two equivalent points on turn is 0.54 nm and is called a pitch. Small or uncharged amino acid residues such as alanine, leucine and phenyl alanine are often found in α-helix. More polar residues such as arginine, glutamate and serine may repel and destabilize α-helix. Proline is never found in α-helix. Hair, nail, skin contain a group of proteins called keratins rich in α-helical structure.

β-pleated sheet structure

A conformation called β-pleated sheet structure is thus formed when hydrogen bonds are formed between the carbonyl oxygens and amide hydrogens of two or more adjacent extended polypeptide chains. Thus the hydrogen bonding in β-pleated sheet structure is interchain. The structure is not absolutely planar but is slightly pleated due to the bond angles. The adjacent chains in β-pleated sheet structure are either parallel or antiparallel, depending on whether the amino to carbonyl peptide linkage of the chains runs in the same or opposite direction.

In both parallel and antiparallel β-pleated sheet structures, the side chains are on opposite sides of the sheet. Generally glycine, serine and alanine are more common to form β-pleated sheet. Proline occurs in β-pleated sheet although it tends to distrupt the sheets by producing links. Silk fibroin, a protein of silk worm is rich is β-pleated sheet.

Tertiary structure

The three dimentional arrangement of protein structure are called a tertiary structurte. The polypeptide chain with secondary structure may be further folded, super-folded, twisted about itself forming many sizes. Such a structural confirmation is called tertiary structure. It is only one such confirmation which is biologically active and protein in this conformation is called as native protein. Thus the tertiary is constitued by steric relationship between the amino acids located far apart but brought closer by folding (**Fig. 37**). The bonds responsible for interaction between groups of amino acids are as follows.

1. **Hydrophobic interactions :** Normally occur between nonpolar side chains of amino acids such as alanine, leucine, methionine, isoleucine and phenyl alanine. They constitute the major stabilzing forces for tertiary structure forming a compact three-dimentional structure.

2. **Hydrogen bonds :** Normally formed by the polar side chains of the amino acids.

3. **Ionic or electrostatic interactions :** The interaction occurs between oppostively charged polar side chains of amino acids, such as basic and acidic amino acids.

4. **van der-Waals forces :** Occurs between non polar side chains.

5. **Disulphide bonds :** These are S-S bonds formed between - SH groups of distant cysteine residues.

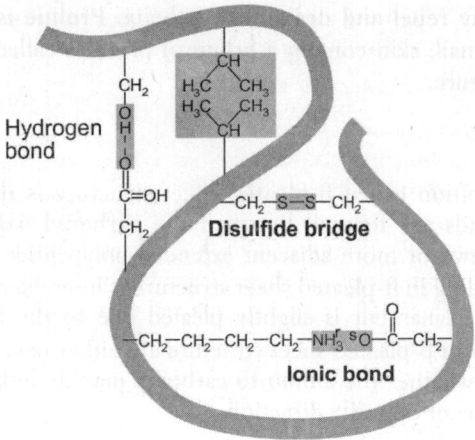

Fig. 37 : Tertiary Structure of Protein.

Domain : Independent folding regions within a protein. The group/pattern of secondary structures forming a Domain's tertiary structure is called a Fold. (Characteristic bond type: hydrophobic; others: hydrogen, ion-pair, van der Waals). A long peptide strand of a protein will often fold into multiple, compact semiindependent folded regions or domains. Each domain having a characteristic spherical geometry with a hydrophobic core and polar surface very much like the tertiary structure of a whole globular protein. The domains of a multidomain protein are often interconnected by a segment of polypeptide chain lacking regular secondary structure. In enzymes with more than one substrate or allosteric effector sites the different binding sites are often located in different domains. In multifunctional proteins, the different domains perform different tasks.

Quarternary structure

Some proteins are made up of more than one polypeptide chain These peptide chains held together by non-covalent interactions or by covalent cross - links it is referred to as the quarternary structure. The assembly is often called as an oligomer and each constituent peptide chain is called as a monomer or sub unit. The monomers of oligomeric protein can be identical or quite different in primary, secondary or tertiary structure.eg: Proteins with 2 monomers (dimer) eg. Creatine phosphokinase Proteins with 4 monomers (tetramer)eg. Haemoglobin. The association of two or more independent proteins via non-covalent forces to give a multimeric protein. The individual peptide units of this protein are referred to as subunits, and they may be identical or different from one another. Proteins that have more than one subunit or polypeptide chains will exhibit quaternary structure **(Fig. 38)**.

Quaternary structure refers to a functional protein aggregate (organization) formed by interpolypeptide linkage of subunits or polypeptide chains. These subunits are held together by noncovalent surface interaction between the polar side chains. Proteins formed like above are termed oligomers and the individual polypeptide chains are variously termed protomers, monomers or subunits. The most common oligomeric proteins contain two or four protomers and are termed dimers or tetramers. respectively. Myoglobin has no quaternary structure

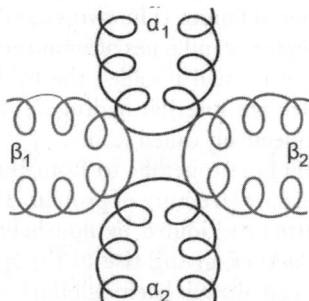

Fig. 38 : *Quaternary Structure of Protein.*

since, it is composed of a single polypeptide chain. Haemoglobin molecule, which consists of four separate polypeptide chains, exhibits quaternary structure. Quaternary structure may influence the activity of enzymes. Some enzymes are active only in their quaternary state and become inactive when split into smaller units. Other enzymes are inactive in the quaternary state and are activated only when they are dissociated to form monomeric state.

Haemoglobin

It is present on red blood cells. It is the vital protein responsible for transport of oxygen and carbon dioxide to and from the body tissues. Haemoglobin is the oxygen-binding protein found in all vertebrates (except fish, belonging to the family Channichthyidae) as well as certain invertebrates. This respiratory protein was first observed in the crystalline form by Friedrich Ludwig Hünefeld, in 1840.

HAEMOGLOBIN MOLECULE

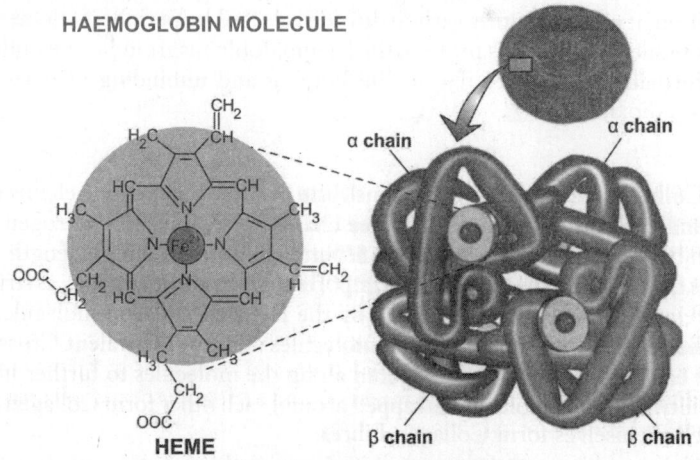

Fig. 39 : *Structure of Haemoglobin.*

Haemoglobin is a globular protein made up of four subunits, each of which contains a polypeptide chain called globin and a heme group. Each haemoglobin molecule is composed of two types of globins organized into four subunits. The two sets of globin chains have minute differences in the sequence and types of amino acids comprising them. This amino acid sequence of a protein is called the primary structure. These amino acid chains fold through internal hydrogen bonding to form helices and sheets, which are collectively called secondary structure of a protein. These secondary structures combine together to form the final three-dimensional structure called the tertiary structure of protein. The globin tertiary structure comprises a helical structures joined by non-helical segments. Four such globins are arranged together, giving rise to the spherical quaternary structure of haemoglobin. Normal adult haemoglobin contains 141 amino acids in each alpha chain, and 146 amino acids in each β chain.

The heme group bound to each globin is an organic macromolecule with an iron at the center. It serves as the prosthetic group of haemoglobin. The prosthetic group of a protein refers to any tightly-bound non-protein entity, that is essential for the structural and functional integrity of the protein. The heme group comprises a structure called the porphyrin ring, which is formed by the combination of four heterocyclic rings called pyrroles. An iron ion (Fe^{2+}) is present at the center of this structure, and is bound to the nitrogen atoms of the four pyrrole rings. It is this central iron which provides the reversible binding to oxygen and carbon dioxide molecules.When the heme is bound to an oxygen molecule or carbon dioxide molecule, it is termed oxyhaemoglobin, or carbaminohaemoglobin respectively. When the heme groups of a haemoglobin molecule are not bound by any molecule, it is referred to as deoxyhaemoglobin. It is oxyhaemoglobin that imparts a bright red colour to blood.

Function : Oxygen transport is the main function of haemoglobin, and more than 98% of the oxygen in blood is carried through haemoglobin. In addition, it also transports carbon dioxide released by peripheral tissues to the lung tissues. During this process, the haemoglobin macromolecule undergoes conformational changes due to the binding and unbinding of oxygen and carbon dioxide.

Collagen

Collagen is a fibrous protein consisting of three polypeptide chains wound around each other. Each of the three chains is a coil itself. Hydrogen bonds form between these coils, which are around 1000 amino acids in length, which gives the structure strength. This is important given collagen's role, as structural protein. This strength is increased by the fact that collagen molecules form further chains with other collagen molecules and form Covalent Cross Links with each other, which are staggered along the molecules to further increase stability. Collagen molecules wrapped around each other form Collagen Fibrils which themselves form Collagen Fibres.

Collagen has many functions. They are as follows. It form the structure of bones, it makes up cartilage and connective tissue, it prevents blood that is being pumped at high pressure from bursting the walls of arteries and it is the main component of tendons, which connect skeletal muscles to bones.

BIOLOGICALLY IMPORTANT PROTEINS

1. Proteins are the essence of life processes.

2. They are the fundamental constituents of all protoplasm and are involved in the structure of a living cell and in its functions.

3. Enzymes are proteins which act as biocatalysts.

4. Proteins serve as components of the tissues holding the skeletal elements together.

5. Nucleoproteins serve as carrier of genetic characters and hence govern inheritance of traits.

6. They execute their activities in the transport of various compounds.

7. Some hormones are proteins which regulate numerous physiological functions.

8. They function in the homeostatic control of the volume of the circulating blood.

9. They are involved in blood clotting through thrombin, fibrinogen and other protein factors.

10. They act as the defense against infections by means of antibodies.

11. Glutathione is a tripeptide containing glutamic acid, cysteine and glycine. It is present in erythrocytes and several other tissues. It acts as a coenzyme and protects haemoglobin against oxidation.

12. Insulin and glucagon are pancreatic hormones, involved in the regulation of glucose metabolism.

13. Angiotensin is a peptide which stimulates the release of certain hormones from adrenal gland.

14. Collagen, is a connective tissue protein rich in proline and hydroxy proline.

15. Interferon (IF or IFN) is a generic term which applies to a number of (about over 20) related low molecular weight, regulatory glycoproteins produced by many eukaryotic cells in response to numerous inducers : a virus infection, double-stranded RNA, endotoxins, antigenic stimuli, mitogenic agents, and many parasitic organisms capable of intracellular growth (Listeria monocytogenes, chlamydiae, rickettsias, protozoa). They are effective in treating viral diseases and cancer, and in eliminating its side effects. The most widely-studied property of interferons is their ability to 'interfere' with the replication of viruses. They are usually species-specific but virus-nonspecific. Interferon was discovered in 1957 in London by the British virologist Alick Issacs and a visiting scientist from Switzerland Jean Lindenmann, both of the National Institute for Medical Research, London.

16.. Peptides from humans called defensins have been found to be antibiotic in nature. Produced by the immune system, these cells smother and kill the invading pathogens. They are secreted by the epithelial cells lining the moist body surface of mammals and serve as the body's own disinfectants.

17. Another group of peptides called endorphins are found in the brain and are involved in the suppression of pain, creation of euphoric highs and feelings of joy.

Introduction

Lipids are a group of organic compounds. They are widely distributed in living organisms. Chemically they are esters of fatty acids. They are capable of forming esters. The important lipids are triacyl glycerols, phospholipids, and sterols.

Fatty acids

Fatty acids are carboxylic acid with hydrocarbon side chains. They are the simplest form of lipids. They are water soluble. They exist in the body either as free acids or fatty acyl esters such as triacylglycerol. The fatty acids are released from lipids on hydrolysis by lipases.

Classification

Fatty acids may be divided into two. They are (i) saturated fatty acids and (ii) unsaturated fatty acids.

Saturated fatty acids

These are fatty acids which donot contain double bonds. They have general formula $CnH_{2n+1}COOH$ (**Table 1**).

Table 1 : **Saturated fatty acids commonly found in natural fats**

Acid	Formula	Carbon atoms
Acetic	CH_3COOH	2
Proplonic	C_2H_5COOH	3
Butyric	C_3H_7COOH	4
Caproic	$C_5H_{11}COOH$	6
Caprylic	$C_7H_{15}COOH$	8
Decanoic	$C_9H_{19}COOH$	10
Lauric	$C_{11}H_{23}COOH$	12
Myristic	$C_{13}H_{27}COOH$	14
Palmitic	$C_{15}H_{31}COOH$	16
Stearic	$C_{17}H_{35}COOH$	18
Arachidic	$C_{19}H_{39}COOH$	20
Behenic	$C_{21}H_{43}COOH$	22
Lignoceric	$C_{23}H_{47}COOH$	24

Unsaturated fatty acid

These are fatty acids which contain double bonds. They have general formula $(C_nH_{2n-1}COOH)$. They are subdivided into

1. Monounsaturated fatty acid : These are fatty acids containing one double bond. (eg) Oleic acid.

$$CH_3(CH_2)_7CH=CH(CH_2)_7COOH$$

Oleic acid

2. **Polyunsaturated fatty acid :** These are fatty acids that contain more than one double bond. (eg) linoleic acid, linolenic acid, arachidonic acid.

Importance

1. They act as energy stores and fuel molecules.

2. They are the major components of cell membrane.

Essential fatty acid (EFA)

The fatty acids that cannot be synthesised by the body and therefore should be supplied in the diet are known as essential fatty acids. Chemically they are polyunsaturated fatty acids (PUFA), namely linoleic acid, linolenic acid and arachidonic acid.

$$CH_3(CH_2)_4CH=CH-CH_2-CH=CH(CH_2)_7COOH$$
Linoleic acid

Functions

1. EFAs are requried for the membrane structure and functions.

2. They are necessary for the maintenance of growth, reproduction and good health.

3. They are important for the transport of cholesterol, formation of lipoprotein and prevention of fatty liver.

4. They serve as precursor for prostaglandin biosynthesis.

5. They prolong clotting time and increase fibrinolytic activity.

Structure of triacyl glycerol

Triacylglycerols are simple lipids in which glycerol backbone is esterified with three fatty acids (**Fig.40**). This form the major part of dietary lipids. They are stored in adipose tissue and serve as concentrated fuel reserve of the body.

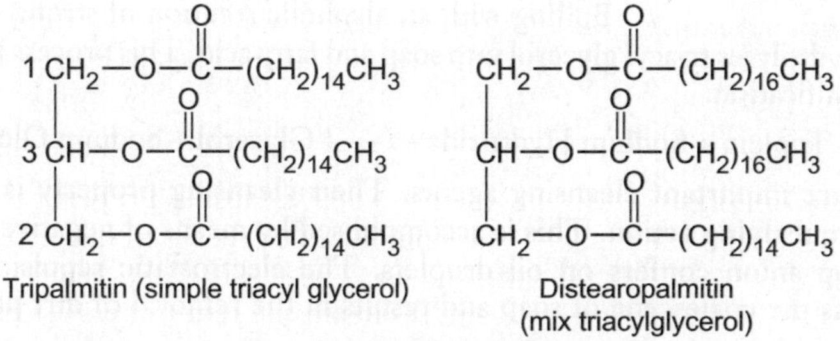

$$1\ CH_2-O-\overset{\overset{O}{\|}}{C}-(CH_2)_{14}CH_3 \qquad CH_2-O-\overset{\overset{O}{\|}}{C}-(CH_2)_{16}CH_3$$

$$3\ CH-O-\overset{\overset{O}{\|}}{C}-(CH_2)_{14}CH_3 \qquad CH-O-\overset{\overset{O}{\|}}{C}-(CH_2)_{16}CH_3$$

$$2\ CH_2-O-\overset{\overset{O}{\|}}{C}-(CH_2)_{14}CH_3 \qquad CH_2-O-\overset{\overset{O}{\|}}{C}-(CH_2)_{14}CH_3$$

Tripalmitin (simple triacyl glycerol)   Distearopalmitin
(mix triacylglycerol)

Fig. 40 : *Triacyl Glycerol*.

If the three hydroxyl groups are esterified with same type of fatty acid then the lipid is called as simple glyceride. If the three hydroxyl groups are esterified with different type of fatty acids, the lipid is called as mixed glyceride.

PROPERTIES

Physical properties

1. Triacylglycerols are non polar, hydrophobic molecules, insoluble in water, but soluble in organic solvents.

2. Specific gravity of fats is lower than water. Therefore fats and oils are float on water.

3. Melting point of triacylglycerol is related to the chain length and degree of unsaturation of fatty acids. The longer the chain length, the higher the melting point and greater the number of double bonds, the lower the melting point.

4. They are tasteless, odourless, colourless and neutral in solution.

5. They are themselves good solvents for other fats.

Chemical properties

1. **Hydrolysis** : On boiling with water at 200°C, triacyl glycerols are hydrolysed to glycerol and fatty acids in a stepwise manner.

Triglyceride $\rightarrow\rightarrow\rightarrow$ Diglyceride+Fatty acid.

Diglyceride $\rightarrow\rightarrow\rightarrow$ Monoglyceride+Fatty acid.

Monoglyceride $\rightarrow\rightarrow\rightarrow$ Glycerol+Fatty acid.

The reaction can also be catalysed by the enzymes lipases.

2. **Hydrogenation** : Hydrogenation of unsaturated fatty acids present in the fats, lead to the formation of saturated fats. Hydrogenation elevates the melting point. Thus oil is converted to a solid fat.

Oleic acid + H_2 $\rightarrow\rightarrow\rightarrow$ Stearic Acid

This reaction is of great commercial importance since it permits transformation of inexpensive and unsaturated liquid vegetable fats into solid fats. The latter are used in the manufacture of candles, vegetable shortenings like vanaspathi and oleomargarine.

3. **Saponification** : Boiling with an alcoholic solution of strong metallic alkali hydrolyses triacyl glycerol into soap and fatty acid. This process is called as saponification.

Triolein + Sodium Hydroxide $\rightarrow\rightarrow\rightarrow$ Glycerol + Sodium Oleate

Soaps are important cleansing agents. Their cleansing property is due to their emulsifying action. This is accomphlised by means of negative charge the soap anion confers on oil droplets. The electrostatic repulsion then prevents the coalescene of soap and results in the removal of dirt particles.

4. **Halogenation** : Unsaturated fatty acids in the triacyl glycerol take up chlorine, bromine and iodine atoms at their double bonds to form saturated halogenated derivatives.

Oleic acid + I_2 $\rightarrow\rightarrow\rightarrow$ 9,10 Diiodo stearic acid

5. **Rancidity** : On storage, unsaturated fatty acids present in the fat are likely to undergo oxidation and hydrolytic cleavage in the presence of lipases present in the fat itself or secreted by the contaminating microorganisms. This leads to change of colour and odour of the fat. This change is called as rancidity. This occurs due to the formation of peroxides at the double bonds of unsaturated fatty acids. Rancidity can be prevented by certain antioxidants such as vitamin-E, gallic acid, butylated hydroxyl toluene etc.Vegetable oils are less rancid because of the presence of natural antioxidants such as vitamin E and carotenoids.

Quantitative tests

There are certain chemical constants used for the characterization of fats.

Acid Number : It is the number of milligrams of KOH required to neutralize the free fatty acids present in 1 gm of fat. The acid number, thus, tells us about the quantity of free fatty acid present in a fat. Obviously, a fat which has been processed and stored properly has a very low acid number.

Saponification number : It is the number of milligrams of KOH requried to saponify 1 gm of fat. The saponification number, thus, provides information on the average chain length of the fatty acids in the fat. The saponification number varies inversely with the chain length of the fatty acids. The shorter the average chain length of the fatty acids, the higher is the saponification number.

Iodine number : It is the number of grams of iodine absorbed by 100 gm of fat. The iodine number is, thus, a measure of the degree of unsaturation of the fatty acids in the fat. The iodine number gives no indication as to the number of double bonds present in the fatty acid molecules.

Polenske number : It is the number of millilitres of 0.1 N KOH required to neutralize the insoluble fatty acids. This indicates the level of non volatile fatty acids present in the fat.

Reichert-Meissl number : It is the number of millilitres of 0.1N KOH required to neutralize the soluble, volatile fatty acids derived from 5 gm of fat. The Reichert-Meissl number thus, measures the quantity of short chain fatty acids in the fat molecule.

Acetyl number : It is the number of milligrams of KOH required to neutralize the acetic acid obtained by saponification of 1 gm of fat after it has been acetylated. The acetyl number is, thus, a measure of the number of OH groups in the fat.

PHOSPHOLIPIDS

Phospholipids are compound lipids containing phosphoric acid in addition to fatty acid, alcohol and a nitrogenous base.

Classification

Phospholipids are classified into two types. They are,

1. Glycerophospholipids (or) Phosphoglycerides that contain glycerol as alcohol.

2. Sphingophospholipids that contain sphingosine as alcohol.

Glycerophospholipids

These are the major lipids that occur in biological membranes. They present in all plant and animal cells. They are abundantly present in heart, brain, kidney, egg yolk and soyabean. The important glycerophospholipids are lecithin, cephalin, phosphotidyl inositol, cardiolipin and plasmalogen.

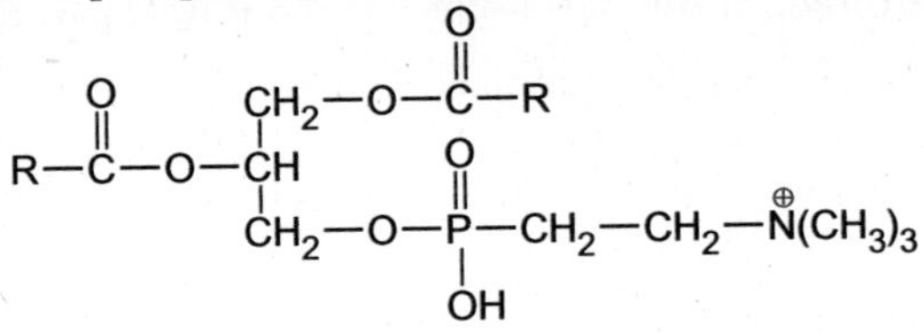

Fig. 41 : **Lecithin.**

The lecithins contain glycerol, fatty acids, phosphoric acid and choline (nitrogenous base). Lecithins generally contain a saturated fatty acid at a1 position and an unsaturated fatty acid at b postition. .

The cephalin contains glycerol, fatty acids, phosphoric acids and ethanol amine as nitrogenous base.

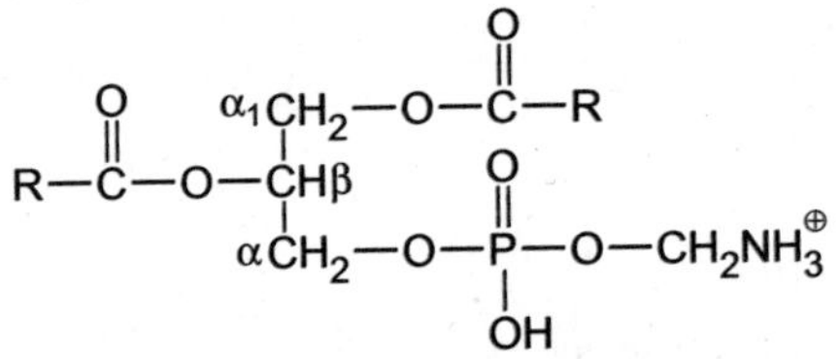

Fig. 42 : **Cephalin.**

Phosphatidyl inositol contains a hexahydric alcohol called as inositol.

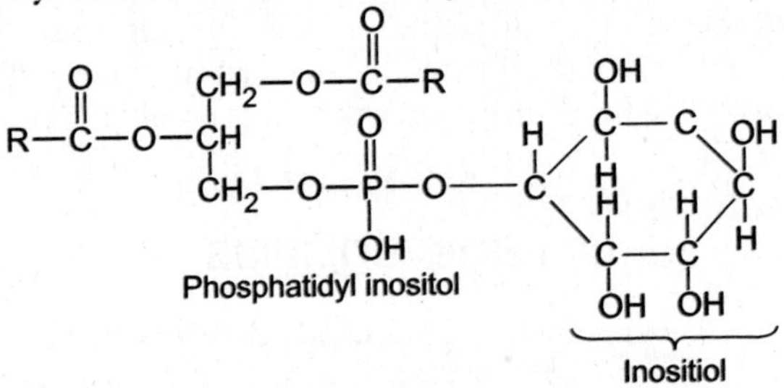

Fig. 43 : **Phosphatidyl inositol.**

Plasmalogens posses an ether link in a1 position instead of ester link. The alkyl radical is an unsaturated alcohol and they are found in brain and nervous tissue.

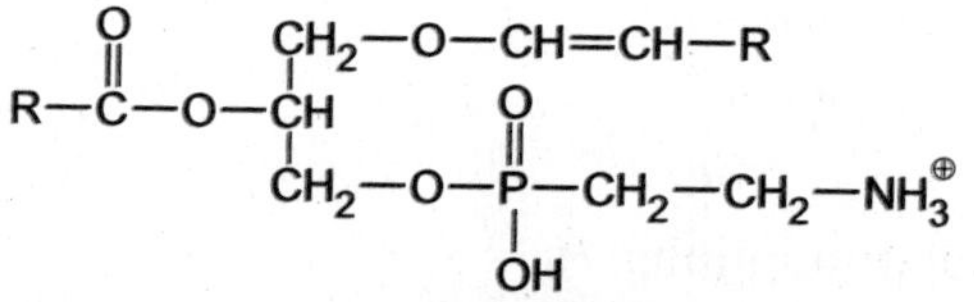

$$R-\overset{\overset{O}{\|}}{C}-O-\underset{\underset{CH_2-O-\underset{\underset{OH}{|}}{\overset{\overset{O}{\|}}{P}}-CH_2-CH_2-NH_3^\oplus}{|}}{\overset{CH_2-O-CH=CH-R}{\underset{|}{CH}}}$$

Plasmalogen

Fig. 44 : **Plasmalogan.**

SPINGOPHOSPHOLIPIDS

These are present in plasma membrane and myelin sheath. They are amphipathic lipids having polar head and non-polar tail. They contain an amino alcohol called shingosine. It is attached to a fatty acid by an amide linkage to form ceramide. Ceramide is linked to phosphoryl choline to form sphingomyelin, which is an important member of sphingophospholipids.

Properties of glycero phospholipids

1. Glycerophospholipids are white waxy substances, which become dark when exposed to air and light, owing to autoxidation and decomposition. This is due to the presence of unsaturated fatty acids in the molecules.

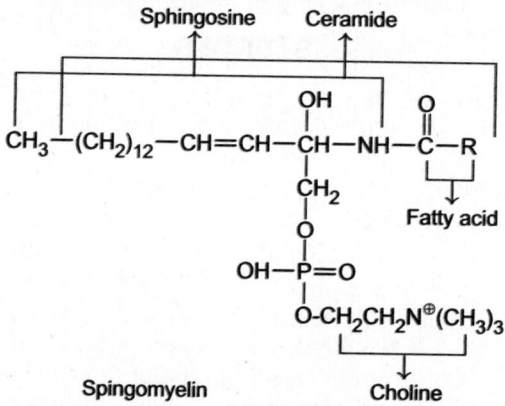

Fig. 45 : **Sphinomyelin.**

2. They are soluble in alcohol and other fat solvents except in acetone.

3. They are hygroscopic and mix well with water to form cloudy, colloidal and slimy solutions.

4. They donot have definite melting point and decompose when heated.

5. They are readily hydrolysed by boiling with acids and alkalies to their constituents.

6. They are hydrolysed by enzyme phospholipase to various component.

Properties of sphingophospholipids

1. They are white crystalline substances

2. They form opelescent suspension in water.

3. They are insoluble in fat solvents like ether and acetone.

4. They are stable in air and light.

Importance of phospholipids

1. They form the structural components of membrane and regulate membrane permeability.

2. They play an important role in cellular respiration.

3. They participate in the absorption of fat from the intestine.

4. They act as surface tension lowering agent.

5. They are essential components of bile where they act as detergents and help in the solubilisation of cholesterol.

6. They also participate in blood clotting.

7. They protect and insulate the neuronal fibres of myelin sheath.

8. They are involved in the interaction of hormones with receptors.

9. They can act as lipotropic agents and prevent fatty liver formation.

10. They help in the reverse transport of cholesterol.

STEROLS

Sterols are compounds containing a cyclic nucleus namely cyclopentanoperhydro phenanthrene (CPPP) and one or more hydroxyl groups. They are widely present in animal and plant tissue.

Cholesterol : Cholesterol is exclusively found in animals and is the most abundant animal sterols. It is widely distributed in all cells and is a major component of cell membrane and lipoproteins. In human beings, it is very important to control the normal level of cholesterol in blood.

Structure : Cholesterol is a $C_{27}(C_{27}H_{46}O)$ compound. It has one hydroxyl group at C_3 and a double bond between C_5 and C_6. An aliphatic side chain is attached to C_{17}. Cholesterol contains a total of 5 methyl groups (**Fig.46**). Cholesterol is the precursor of various physiologically important compounds such as bile acids, vitamin-D, steroid hormones etc.

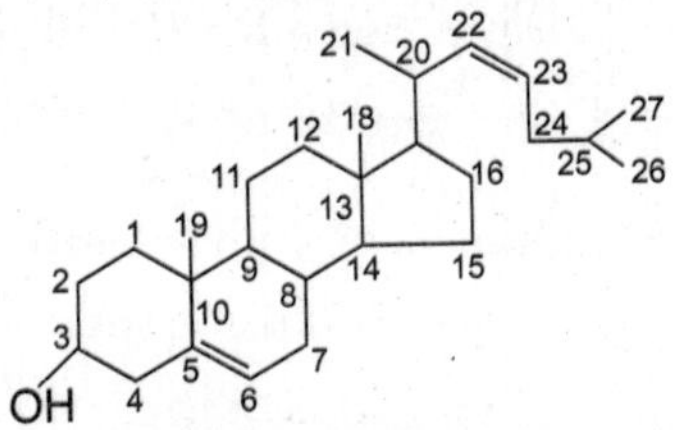

Fig. 46 : Cholesterol.

Properties

1. They are white shining rhombic plate like crystals.

2. It is tasteless and odourless

3. It has a high melling point of 150°C.

4. It is insoluble in water and soluble in fat solvents.

5. It is a poor conductor of heat and electricity and serves as an insulator against electric charge. In brain, where it is present abundantly, it acts as an insulator against nerve impulse which are electrical in nature.

6. Cholesterol, when oxidised under suitable conditions, undergoes rapid oxidation to form a ketone-cholestenone.

7. The hydroxyl group of cholesterol readily forms ester with fatty acids, stearic acid etc.

8. It gives addition reactions such as hydrogenation and halogenations because of the presence of double bond.

Physiological importance of cholesterol

1. It is one of the essential constituents of cells.

2. It influences the permeability functions of the cell.

3. It controls the redcells from being easily hemolyzed.

4. It performs defensive action.

5. It assists the formation of bile acids and bile salts, 7- dehydrocholesterol, vitamin D_3, corticosteroid hormones, androgens, estrogens and progesterone.

6. It acts as an antagonist to phospholipid.

ERGOSTEROL

Ergosterol occur in plants. It is also found in yeast and fungi as the structural constituent of membranes. It is an important precursor for vitamin-D. When exposed to light, it is converted to ergocaliciferol, a compound containing vitamin-D activity. Its structure is similar to that of cholesterol, but differs in the following aspects.

1. It has double bond at C_7-C_8.
2. It has a double bond in the side chain.
3. It has an additional CH_3 group in the side chain (**Fig.47**).

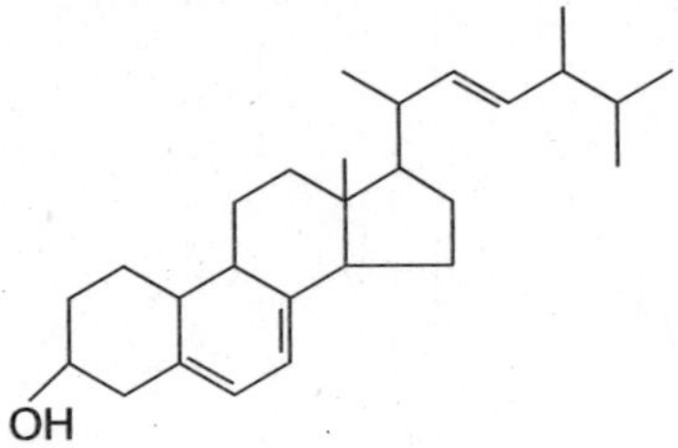

*Fig. 47 : **Ergosterol.***

Stigma sterol - **It is structurally similar to that of ergosterol except at C$_7$ (Fig.48).**

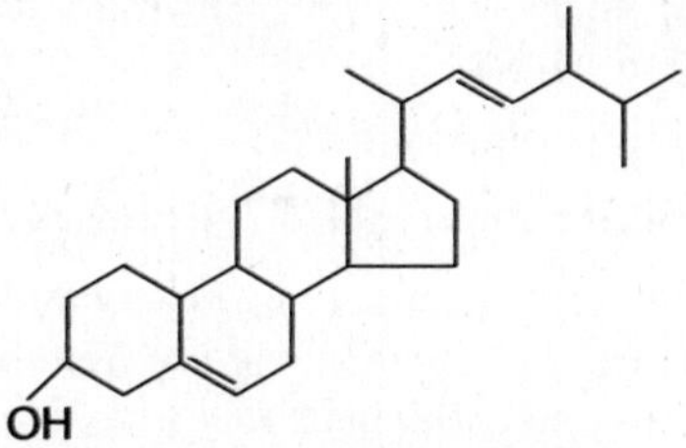

Fig. 48 : **Stigma sterol.**

Stigmasterol and its derivatives sitosterols are probably the most common sterol of plants. The important sources are soya bean and calabar beans.

Introduction

Nucleic acids are high molecular weight compounds. Nucleic acids are the chemical basis of life and heredity. They are found in all living cells. It is found inside of the nuclei of eucaryotic cells. It was first isolated in 1868 by Johann Friedrich Miescher from the nucleus. He called this as nuclein. They contains purine and pyrimidine bases, sugar and phosphoric acid. Nucleic acids can be divided into two main classes depending on the sugar they contain: deoxyribonucleic acids (DNA) contain 2-deoxy-d-ribose and ribonucleic acids (RNA) contain d-ribose. Nucleic acids are colourless, complex, amorphous compounds. Fischer in 1880 indicated that nuclein contain purines and pyrimidines. Altmann in 1889 gave the term nucleic acid. Nucleic acids serves as transmitters of genetic information.

Structure

Nucleic acids are the polymers of nucleotides held by 3′ and 5′-phosphate bridges. The nucleotide consists of nucleobase, sugar and phosphate. The term nucleoside refers to nucleobase and sugar. Nucleotide refers to nucleoside and phosphate. The nitrogenous base found in nucleotides are purines and pyrimidines.

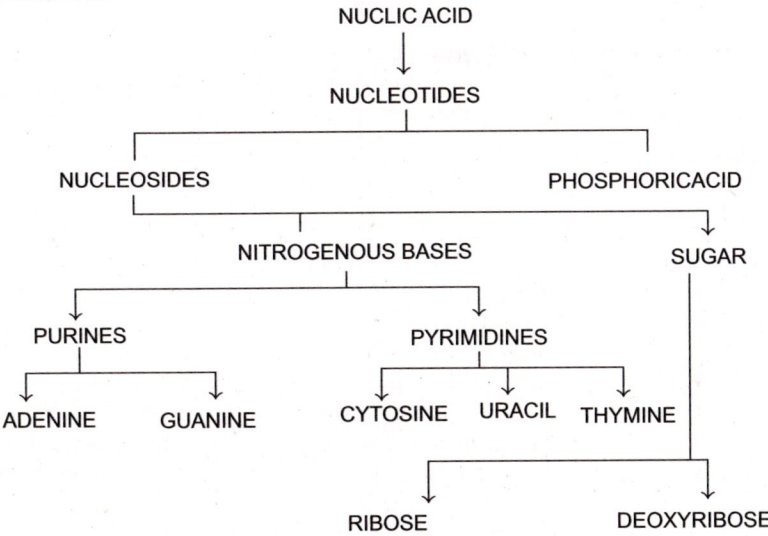

*Fig. 49 : **Nucleic acid-Components**.*

STRUCTURAL COMPONENTS OF NUCLEIC ACIDS

Phosphoric acid

The molecular formula of phosphoric acid is H_3PO_4. It contains 3 monovalent hydroxyl groups and a divalent oxygen atom, all linked to a pentavalent phosphorus atom.

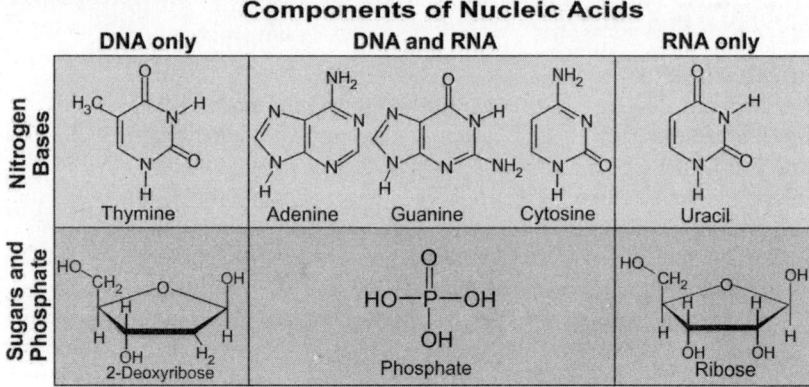

Fig. 50 : *Components of Nucleic acid.*

Pentose Sugar

The two types of nucleic acids (DNA and RNA) are distinguished primarily on the basis of the 5 carbon sugar pentose which they possess. One possesses deoxyribose, (deoxyribonucleic acid) while the other contains D-ribose (hence called ribonucleic acid). Both these sugars in nucleic acids are present in the furanose form and are of α configuration.

Nitrogenous Bases

Two types of nitrogenous bases are found in all nucleic acids. These are derivatives of purine and pyrimidine.

Purine

Purine are aromatic organic compound that consists of a **pyrimidine** ring and an imadazole ring ($C_3H_4N_2$). Eg. Adenine and Guanine. Purine bases are derived from the parent compound purine. Purine contains the heterocyclic ring system. Fusion of the pyrimidine ring with imidazole yields the purine ring. The purine bases present in nucleic acids are adenine and guanine. Other purine bases are hypoxanthine and xanthine. They are intermediates in the formation of adenine and guanine nucleotides. Purine bases are sparingly soluble in water. They absorb light in UV region at 260 nm. This property is used for the detection of and the quantification of purine nucleotides. They are capable of forming hydrogen bonds. They exhibit keto-enol tautomerism at body pH.

Pyrimidine

Pyrimidine is a aromatic organic compound with 2 nitrogens at C_1 and C_3 of a six membered ring. Pyrimidine bases are derived from the parent compound pyrimidine. The pyrimidine bases present in nucleotides are cytosine, uracil and thymine. It is 6 membered heterocyclic compound contains carbon, nitrogen, hydrogen and oxygen atoms. Pyrimidine bases are soluble in water. They absorb UV light at 260 nm. This property is used to detect and estimate pyrimidine nucleotides. They are capable of forming hydrogen bonds with other purine bases. They exhibit keto-enol tautormerism.

Nucleosides

A nucleoside is composed of purine or pyrimidine base and a pentose sugar. Two types of pentose sugar are present in nucleoside, they are ribose and deoxy ribose. In the case of purine nucleosides, the sugar is attached to N-9 of the purine ring, whereas in pyrimidine nucleosides, the sugar is attached to N-1 of the pyrimidine ring. The type of linkage is N-glycosidic linkage.

Nucleotides

Nucleotides are phosphorylated nucleosides usually one or two hydroxyl groups of ribose (or) deoxyribose and are phosphorylated. Thus a nucleotide has three structural components. They are nitrogenous base, sugar and phosphate. Phosphate is attached to ribose (or) deoxy ribose through an ester linkage.

Nucleotides are the fundamental units of nucleic acids. Each nucleotide is comprising of a

(a) Phosphate group, (b) Pentose Sugar and (c) Nitrogenous base.

Base Pairing

Base pairing is an essential feature not only to maintain the double helical structure of DNA, but also plays an important role in DNA, RNA and protein biosynthesis.

In DNA

Adenine (A) pairs with Thymine (T) (A=T)
Guanine (G) pairs with Cytosine (C) (G≡C)

In RNA

Adenine (A) pairs with uracil (U) (A=U)
Guanine (G) pairs with cytosine (C) (G≡C)

Primary Structure of Nucleic Acids

The sequence or order of the nucleotides defines the primary structure of DNA and RNA. The nucleotides of the polymer are linked by phosphodiester bonds connecting through the oxygen on the 5′ carbon of one to the oxygen on the 3′ carbon of another. The Oxygen and Nitrogen atoms in the backbone give DNA and RNA.

Types of Nucleic Acids

There are two types of nucleic acids. They are deoxyribonucleic acid or DNA, and ribonucleic acid or RNA.

DEOXYRIBO NUCLEIC ACID

DNA is the hereditary material. It is found in the nucleus and mitochondria of a cell. DNA is the principal genetic material of all known living organisms. DNA is chemically called a nucleic acid. It is found in all cells except few viruses. In prokaryotes, DNA exist as nucleoid in the cytoplasm. In Eucaryotes, it is available within a nucleus. Nucleotide sequence of a nucleic acid is known as its primary structure which confers individuality to the polynucleotide chain. Polynucleotide chain has direction. They are represented in $5' \rightarrow 3'$ and $3' \rightarrow 5'$ directions. Each polynucleotide chain has 2 ends. The $5'$ end carrys a phosphate group and $3'$ end carrying an unreacted hydroxyl group. The DNA is highly fragile. It is determined by its length. Larger DNA breaks easily. High temperature, acid or alkali treatment denatures the DNA and a denatured DNA may gets its original nature by correcting the environment. DNA plays a major role in all biosynthetic and hereditary functions of all living organism thus they carry genetic information from one generation to other. DNA is a stable macro molecule and it is immortal. It controls all developmental process of an organism. It synthesis RNA. Its genetic code is responsible for protein synthesis.

Double helix structure

In 1953, J.D. Watson and F.H.C. Crick proposed a precise three dimensional model of DNA structure based on model building studies, base composition and X-ray diffraction studies. This model is popularly known as the DNA double helix. They won nobel prize in 1962. DNA is made up of two polynucleotide chains. Each polynucleotide contains three chemical components, they are phosphoric acid, deoxy ribose sugar and a nitrogen base. Nitrogen bases are two ringed purines like adenine (A) and guanine (G) and two or one ringed pyrimidine cytosine (C) and thymine (T). The DNA contains 4 types of nucleotides namely AMP (Adenosine monophosphate), GMP (Guanosine monophosphate) TMP (Thymine monophosphate) and CMP (Cytocine monophosphate).

Nucleotide Composition

Base	Ribonucleoside (Base + Ribose)	Ribonucleotide (Base + Ribose + Phosphate)
Adenine (A)	Adenosine	Adenosine 5'-monophosphate (AMP)
Guanine (G)	Guanosine	Guanosine 5'-monophosphate (GMP)
Cytosine (C)	Cytidine	Cytidine 5'-monophosphate (CMP)
Uracil (U)	Uridine	Uridine 5'-monophosphate (UMP)

In each nucleotide, the deoxyribose sugar is attached to a phosphoric acid at one side and a nitrogen base at the other side. The nitrogen base is joined to the sugar by a glycosidic bond.

Two nucleotides are joined by phosphodiester bond. Many nucleotide are linked together to form a polynucleotide chain. The nucleotide of adjacent chains are linked by hydrogen bond.

A purine base always pairs with a pyrimidine base or more specifically Guanosine (G) with Cytosine (C) and Adenine (A) with Thymine (T) or Uracil (U). Adenine always linked with thymine (A=T). Similarly guamine is linked with cytocine (G $\rightarrow$ C).

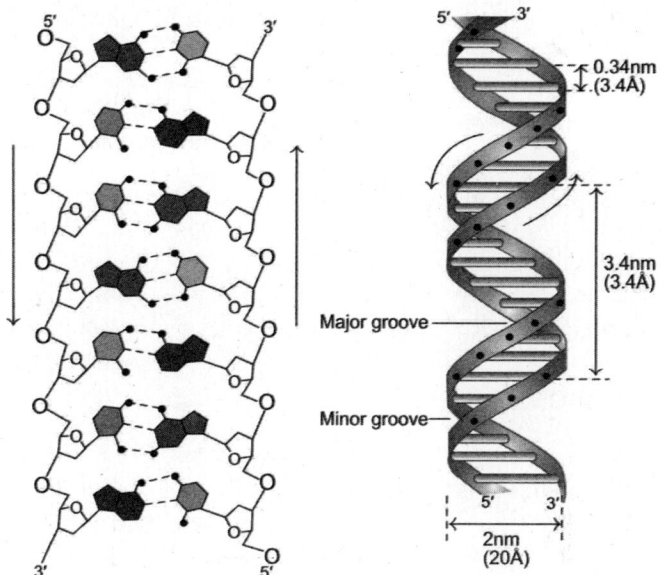

*Fig. 51 : **Double Helix Structure and DNA.***

The amount of adenine is equivalent the amount of thymine and the amount of guamine is equivalent to the amount of cytosine. This is called Chargalf's equivalent rule (1950).

The two chains of DNA are complementary to each other. The two polynucleotide chains of DNA molecules are coiled around each other to form a double helix.

The helix has a diameter of 20A° between the two complementary strands and makes one complete turn for every 34A°: Hence each turn contains 10 nucleotides.

The DNA helix has two grooves, namely major groove (deep wide) and minor groove (Shallow narrow).

The two strands run anti parraley to each other.

One strand has phosphodiester linkage in $3' \rightarrow 5'$ direction while other strand has phosphodiester $5' \rightarrow 3'$ direction. Size of the DNA molecules varies from organisms to organisms.

Chargaff's rule of DNA composition

DNA has equal number of adenine and thymine residues (A=T) and equal number of guanine and cytosine (G=C) residues. This is known as Chargaffs rule of molar equivalence between purine and pyrimidines in DNA.

Types of DNA

DNA is classified in to types in different ways. DNA is classified into two based on the number of strands. They are Double Stranded (DS) DNA and Single Stranded (SS) DNA. DS DNA : It is a Watson and cricks double helix DNA. SS DNA : Some viruses have SS DNA.

DNA is classified into three on the basis of number of nucleotide residues present in each turn of the helix. They are A DNA, B DNA and Z DNA.

A DNA : It is a double helical DNA have 11 residues per turn. It is A° in diameter. It has a right handed helix. It is formed by dehydration of B DNA.

B DNA : It is a biologically important DNA. It is commonly & naturally found in most living system. It has 10 residues per turn of the helix. It is also a right handed helix. It is the Waston and Crick double helix.

Z DNA : It is a left handed double helix having 12 residues per turn.

On the basis of the shape, DNA is classified into 3 types namely circular DNA, Relaxed DNA and super coiled DNA. Circular DNA is circular in shape, eg., ϕX174. Relaxed DNA is without any helical coiling. Supercoiled DNA is a circular helix and is twisted form of a super helix.

Functions of DNA

1. DNA is the genetic material of living organisms.

2. DNA contains all the information required for the functioning of an individual organism.

3. The genetic information in DNA is converted to characteristic features of living organisms like colour of the skin and eye, height, intelligence, ability to metabolise particular substance, ability to withstand stress, susceptibility to diseases and ability to produce certain substances.

4. DNA is the source of information for the synthesis of all cellular proteins. The segment of DNA that contains information for a protein is known as gene.

5. DNA is transmitted from parents to offsprings and hence transfers genetic information from one generation to another.

RNA

RNAs is present in the nucleus, ribosomes and cytoplasm of eukaryolic cells. They are involved in the transfer and expression of genetic information. They act as primer for DNA formation. Some act as enzymes and as coenzymes. RNA also functions as genetic material in some viruses. RNAs are also polynucleotides. In RNA polymer, purine and pyrimidine nucleotides are linked together through phosphodiester linkages. The sugar present is ribose. The nitrogenous bases present in RNA are adenine and guanine (purine bases), uracil and cytosine (pyrimidine bases). The nucleotides present in RNA are adenylic acid, quanidylic acid, cytidylic acid and uridylic acid.

RNA has several functions and is found in the nucleus, cytosol and mitochondria. Messenger RNA (mRNA) carries genetic information obtained from DNA to sites that translate the information into a protein. Transfer RNA (tRNA) carries activated amino acids to sites where the amino acids are linked together to form polypeptides. Ribosomal RNA (rRNA) is a structural component of ribosomes, which serve as the sites for protein synthesis. Small nuclear RNA (snRNA) is a component of small nuclear ribonucleoprotein particles. These particles process heterogeneous RNA (hnRNA, the immature form of mRNA) into mature mRNA. In some viruses, HIV, influenza, polio, RNA functions as the storage house of genetic information.

TYPES OF RNA

There are mainly three types of RNA s in all prokaryotic and eukaryotic cells. They are (1) Messenger RNA (mRNA) (2) Transfer RNA (tRNA) (3) Ribosomal RNA (rRNA). They differ from each other by size, formation and stability and importantly with function.

1. **Messenger RNA** : The term mRNA was coined by Jacob and Monad in 1961. It accounts for 1-5% of cellular RNA. They are single stranded linear molecules and carry genetic message in the form of triplet codon. They consist of 1000-10,000 nucleotides.

They have a free or phosphorylated 3′ and 5′ end. They have different life span ranging from few minutes to days. mRNA molecules are capped at 5′ end by methylated guanine triphosphate. It is used to bind ribosome. Capping protects mRNA from nuclease attack. The cap is followed by non coding region (Fig.52).

This is followed by initiation codon, coding region, termination codon and noncoding region. At 3′ end, a polymer of adenylate (poly A) is found as the tail in eucaryotes. Poly A tail protects mRNA from nuclease attack. Intrastrand base pairing among complementary bases allows folding of the linear molecule. As a result, haripin or loop like secondary structure is formed. Prokaryotic mRNA are polycistronic in nature. The mRNA in nucleus are called hnRNA (heterogenous nuclear RNA). The mRNA is synthesed from DNA by RNA polymerase and the process called transcription. Each mRNA mostly contains the codons for one polypeptide chain although differences do exist Procaryotic mRNA are called Polycistronic. It contains several sites for initiation and termination of polypeptide due to the presence or sevaral structural genes. Therefore one mRNA synthesis more than one poly peptide.

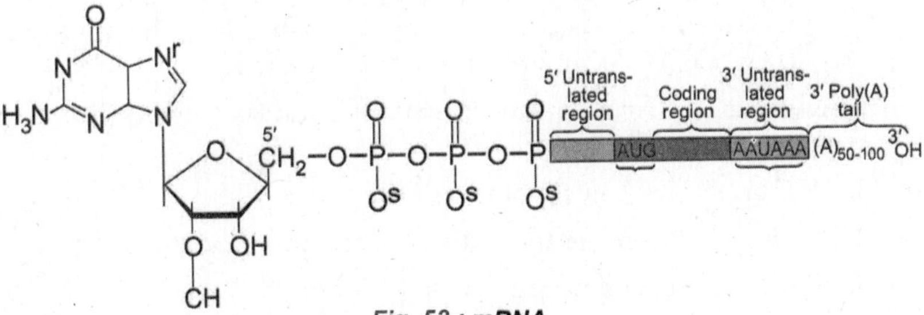

Fig. 52 : mRNA.

Functions

1. mRNA is the direct carrier of genetic information from the DNA in the nucleus to the cytoplasm.

2. It contains information required for the synthesis of protein molecules.

TRANSFER RNA

It transfers the activated amino acids to the ribosome. It is also called as soluble RNA, supernatant RNA and adapter RNA. It accounts for 10-15% of total cell RNA. They are the smallest of all the RNAs. Usually they consist of 50-100 nucleotides. They are single standard molecules. They contain unusual bases such as methylated adenine, guanine, cytosine and thymine, dihydrouracil and pseudouridine. These unusual bases are important for binding RNA to intra chain base pairing. Further some bases are not involved in base pairing, resulting in loops and arms formation in tRNA. These folding in the primary structure generates a secondary structure (**Fig.53**).

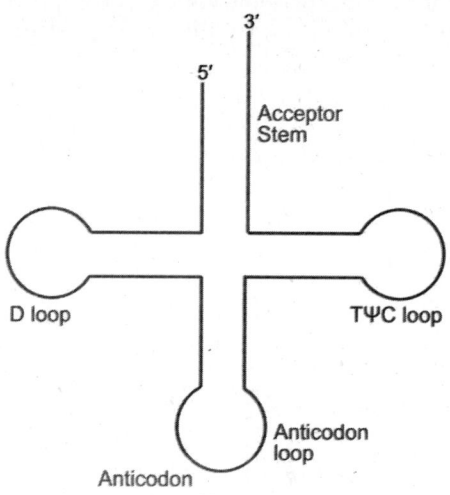

*Fig. 53 : **tRNA.***

Secondary structure of t-RNA is in the form of a clover leaf. The important feature of the clover-leaf structure are,

1. An amino acid acceptor arm does not end with a loop. It has constant base sequence "CCA" 3'-OH of adenosine moiety of t-RNA.

2. An anticodon arm, stem has 5 paired bases and the loop has 7 unpaired bases. It recognises codon on mRNA.

3. TΨC arm which contains unusual base cytosine.

4. D- arm which contains many dihydrouracil residues.

5. Variable arm has miniature stem only present.

Functions

1. It is the carrier of amino acids to the site of protein synthesis.

2. There is at least one t-RNA molecule to each of 20 amino acids required for protein synthesis.

RIBOSOMAL RNA

It is a single stranded RNA. It is a polynucleotide chain. This accounts for 80% of the total cellular RNA. It is present in ribosomes. In ribosomes, r-RNA is found in combination with protein. It is known as ribonucleoprotein. The length of rRNA ranges from 100-600 nucleotides. rRNA molecules have a secondary structure. Intra strand base pairing between complementary bases generate double helical segments or loops based on sedimentation coefficient. rRNA are classified into 7 types. They are 28S, 18S, 5.8, 5S (Eucaryotic ribosomal RNA), 23S, 16S and 55S (Procaryotic **ribosomal RNA**).

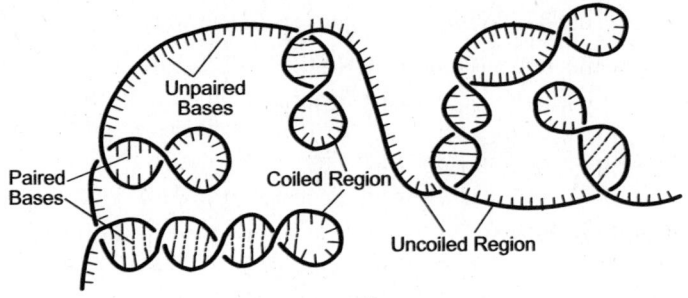

Fig. 54 : rRNA.

Functions

(a) rRNAs bind protein molecules and give rise to ribosomes, (b) 32 end of 18S rRNA (16S in procaryotes) has nucleotides complementary to those of cap region of mRNA. (c) 5S rRNA and surrounding protein complex provide binding site for tRNA. (d) rRNAs get associated with specific proteins to form ribosome subunits. 50S subunit of prokaryotic ribosome contains 23S rRNA, 5S rRNA and some 32 protein molecules.

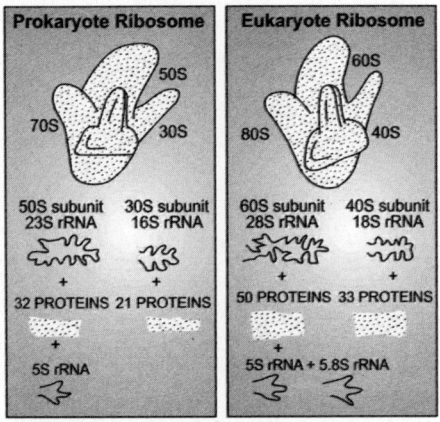

*Fig. 55 : **Generalized structure of ribosome in prokaryotes and eukaryotes.***

30S subunit of prokaryotic ribosome has 16S rRNA and about 21 protein molecules. 60S subunit of eukaryotic ribosome contains 28S rRNA, 5S rRNA, 5.8S rRNA and about 50 pro-tein molecules. 40S subunit of eu-karyotic ribosome consists of 18S rRNA and some 33 protein molecules

In eucaryotic transcript the arrangement in $5' \to 3'$ direction is 18S–5.8S–28S. Several methylations occur prior to removal of spacer RNA. Removal of spacer RNA breaks the transcript into 2-3 parts. 5S is often transcribed separately.

They are required for the formation of ribosomes.

They are involved in the initiation of protein synthesis.

4. Genomic RNA (Genetic RNA): It is found in some viruses called riboviruses. Ge-nomic RNA may be single stranded (e.g., Tobacco Mosaic Virus or TMV) or double stranded (e.g., Reovirus). It is fragmented in influenza virus. Genomic RNA acts as a hereditary material. It may replicate directly, or form DNA in the host cell to produce RNA of its own type.

5. Catalytic RNAs: Cech et al (1981) found catalytic activity (cleavage and covalent bond formation) in RNA precursor of ciliated protozoan called Tetrahymena thermophila. It was called ribozyme. In 1983, Altman et al discovered that ribonuclease – P that takes part in processing tRNA from its precursor is a biocatalyst made of RNA and protein. Noller et al (1992) found peptidyl transferase to be RNA enzyme.

6. Small Nuclear RNA (snRNA): It is a small sized RNA present in the nucleus. Each RNA is combined with 7-8 molecules of proteins to form small nuclear ribonucleoprotein or snRNP. SnRNA takes part in splicing (U1 and U2), rRNA processing (U3) and mRNA processing (U7).

7. Small Cytoplasmic RNA (scRNA): It is small sized RNA occurring free in the cytoplasm. One such small cytoplasmic RNA is 7S and combines with 6 protein molecules to produce signal recognition particle or SRP. The latter helps in taking and binding a ribosome to endoplasmic reticulum for producing secretory proteins.

Introduction

The vitamins are a complex organic compounds. They are required in the diet in small quantities to perform specific biological activities. It is required by the body for the maintenance of good health. The vitamins are present in foods in small quantities. Each vitamin has a specific chemical structure and a specific function. Most of the vitamins act as coenzymes in the body. Normally a well balanced diet will supply all the necessary vitamins in sufficient quantity. Funk in 1917 coined the term Vitamine means Vital for Life.

Plants synthesis vitamins efficiently. Vitamins are not synthesized in the mammalian body. Most of the bacteria are able to synthesis vitamins. For human it should be supplied through the diet. Provitamin is an organic factor which is present in food. Provitamin is converted into vitamin. It is a precursor of vitamin. Eg. Carotene is a provitamin of Vitamin A.

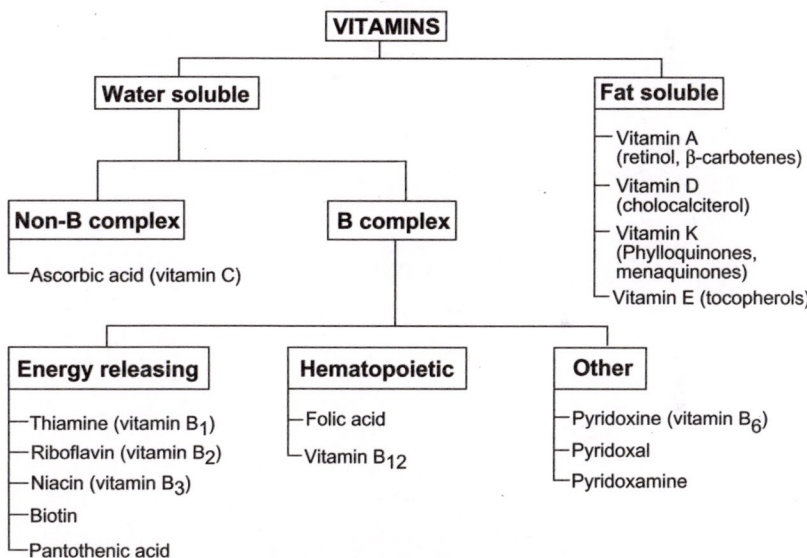

Fig. 56 : *Classification of vitamins.*

Classification

Vitamins are generally classified into two main groups. They are (i) fat soluble vitamins (ii) water soluble vitamins (**Fig.56**).

Fat soluble vitamins

These are stored in adipose tissues. All these vitamins are not soluble in water and readily soluble in fat dissolving organic solvents. Inside the body, these vitamins need fat for their transport and metabolism.. Eg. A, D, E and K.

Water soluble vitamins

They are readily soluble in water. Eg., B complex vitamins and vitamin C. B-Complex vitamins are again subdivided into energy releasing vitamins, haematopoietic vitamins and other. .

Vitamin A

It was discovered by Mc Collum in 1913. It is a fat soluble vitamin. The yellow plant pigments α,β and γ-carotenes and cryptoxanthin are precursors of vitamin A. The body has the ability to convert these carotenoid compounds present in the diet into vitamin A. Vitamin A is found only in foods of animal origin. It is present in almost all species of fish, birds and mammals. There are two forms of vitamin A. They are Vitamin A_1 which occurs in the liver of marine water fish and Vitamin A_2 found in the liver of fresh water fish. The vitamin A which contain alcoholic group in the side chain is called as retinol and which contain aldehyde group is known as retinal. Though the two vitamins differ slightly in their chemical structures their physiological functions are the same.

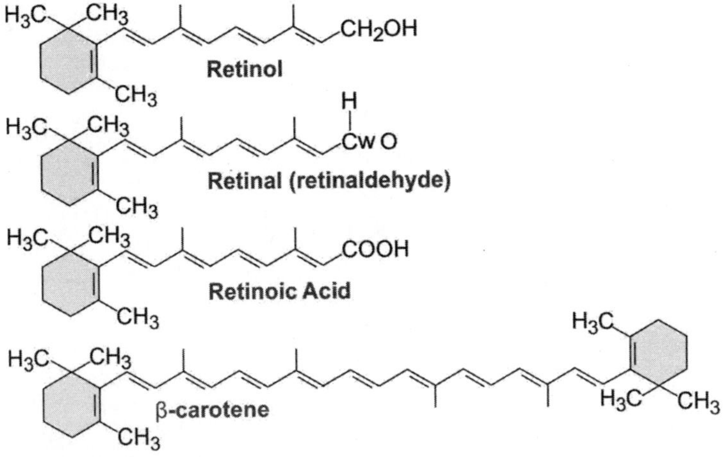

Fig. 57 : **Vitamin A.**

Retinol is a primary alcohol with β-ionone ring.

Retinal is an aldehyde form obtained from retinol by oxidation.

Retinoic acid is a vitamin A acid. It is obtained from retinal by oxidation. β-carotene from the plant food are hydrolysed in our body and produce retinal by a specific enzyme β-carotene 15,15 dioxygenase in the intestine and it is reduced by retinol by dehydrogenase. In the intestinal mucosal cells, retinol is reesterified to long chain fatty acids and transported to the lymph. Then retinol esters are taken up by liver and stored. As and when needed, vitamin A is released from liver as free retinol. Retinol is released into the circulation by plasma retinol binding protein with the aid of zinc. Many cells contain cellular retinol binding protein, which carried retinol to the nucleus and binds to the chromatin.

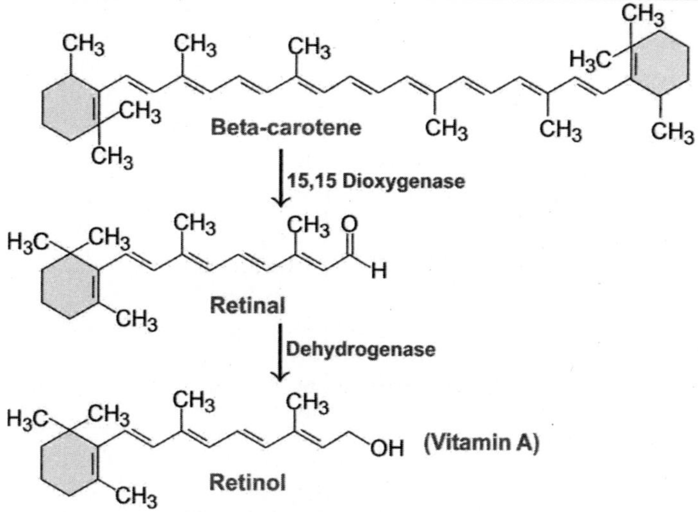

*Fig. 58 : **Metabolism of β carotine.***

Functions

Vitamin A is essential

1. for the growth and metabolism of all body cells.

2. for the formation of rhodopsin (visual purple), a complex substance formed from retinol and protein.

3. for the maintenance of healthy skin, particularly mucous membrane of the cornea and the lining of respiratory tract.

4. It is essential for protein synthesis.

Visual Cycle

Rhodopsin, a pigment found in retina. It is necessary for vision in dim light. The biochemical process of Vitamin A in vision was first elucidated by George Wald(Nobel Prize 1968). The event occurs in cyclic process and called Rhodopsin cycle or Walds visual cycle. Visual cycle means biological conversion of a photon into an electrical signal in retina. This conversion taken place in the photoreceptor cells. Rods and cones are the major part of photoreceptors cells. Rhodopsin is a conjugated proteins present in rods. Rhodopsin is also called visual purple. It contains scotopsin and retinal (as 11-Cis retinol). On exposure to light, the colour of rhodopsin changes from red to yellow by a process called bleaching. During this process there is a isomeration takes place between 11-cis retinal and all trans retinal, which is responsible for reformation of rhodopsin.

Cones are specialized in bright and colour vision. Colour visoion is governed by 3 pigments like porphyropsin (red), iodopsin (Green) and Cyanopsin (Blue). All these are available in retinal–Opsin complex. These pigments become bleached, when light strikes cones and rods. This leads to conformational change in opsin. This opsin is responsible for the generation of nerve impulse. This causes perception of specific colour.

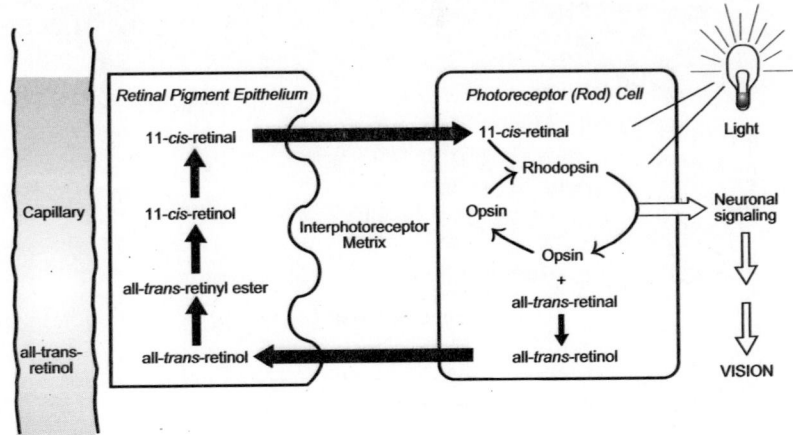

*Fig. 59 : **Visual Cycle.***

Process of bleaching-Light-Rhodopsin–Parthorhodopsin-lumirhodopsin-metarhodopsin I-metarhodopsin II-all Trans retinal-11-Cis retinal-opsin-neurosignalling-light.

Sources : The liver of any animal is a rich source of vitamin A. Fish liver oil is an excellent source of vitamin A. Whole milk, egg yolk, dark green leafy vegetables and deep yellow vegetables and fruits are rich in carotenes, which can be converted into vitamin A by the intestinal wall.

Requirements

Vitamin A requirement is based on the intake to maintain the normal blood level. The capacity of the body to store vitamin A provides for an effective emergency supply. Recommended amount of Vitamin A for different age group is as follows:

Infants: 1500 IU/day, Children: 2000-3000 IU/day, Adults: 5000 IU/day.

Pregnant and lactating women - 6000–8000 IU / day (IU = International units)

Deficiency

Night blindness. It is the earliest sign of vitamin A deficiency. The individuals have difficulty to see in dim light due to increased dark adaptation time.

More prolonged or severe deficiency leads to the ulceration of cornea and this condition is known as **xerophthalmia or keratomalacia**.

Vitamin A deficiency results in growth retardation due to impairment of skeletal formation.

The reproductive system is adversely affected in vitamin A deficiency.

The skin becomes rough and dry.

Vitamin D

It is a fat soluble vitamin. It resembles steroid in its structure and function. There are two distinct forms of Vitamin D. Cholecalciferol (Vitamin D3) is the natural form of the vitamin occurring in foods. It can be formed under the skin from 7 dehydro cholesterol by the influence of sun light (ultra violet (UV) radiation). Ergo calciferol (vitamin D2) is a synthetic form of the vitamin D which has the same activity as the natural vitamin. It is produced by the UV radiation of ergo sterol, a compound which can be extracted from yeast. This is the form of vitamin which is added to commodities such as margarine and baby foods.

Functions : Vitamin D is necessary for the growth and maintenance of bones and teeth. Vitamin D helps in the absorption of calcium and phosphorus from the small intestine and for the uptake of calcium and phosphorus by the bones and teeth.

Sources : Vitamin D is not widely distributed in nature and the best sources are fish oil, especially liver oil. Milk, butter and egg yolk are the only foods in the ordinary diet that contain vitamin D.

Requirements

It is difficult to make standard recommendations for the dietary requirements of vitamin D since the amount of vitamin D produced in the body by the action of sunlight varies from person to person. Many people may obtain this vitamin from sunlight. It is, however certain that babies and growing children require more vitamin D than adults, due to rapid growth and bone development.

Infants and children - 400 IU / day

Adults - 200 IU / day

Pregnant and lactating women - 400 IU/day

Absorption and storage : Fat helps in the absorption of vitamin D and bile is essential for its absorption. Vitamin D enters into the general circulation via lymph and stored largely in liver and kidneys.

Deficiency : Children receiving an inadequate supply of vitamin D develop **rickets**. Calcium and phosphorus are inadequately deposited in the bones. Premature infants are more susceptible to rickets than full term infants. In adults on inadequate supply of vitamin D causes **osteomalacia**, a condition in which the bones become soft, weak and painful.

Vitamin E

It was discovered by Matill and Conclin in 1920. It is a fat soluble vitamin. It is a naturally occurring antioxidant. It is also called as anti sterility vitamin as this is essentials for reproduction in animals. Vitamin E activity is possessed by a number of compounds known as tocopherols. In 1936, Evans and his collegues isolated a compound with vitamin E activity and named as tocopherol (ingreak tokos - Child birth; Pher-to beal; ol-alcohol). About eight tocopherols are identified. They are called as $\alpha,\beta,\gamma,\delta$ etc., Among these compounds α-tocopherol is known as vitamin E and more active. Structurally vitamin E contains a dihydro benzopyran nucleus with an isoprenoid side chain. Many of them are pale yellow oils soluble in fat. Embrical formula of α-tocopherol is $C_{29}H_{50}O_2$.

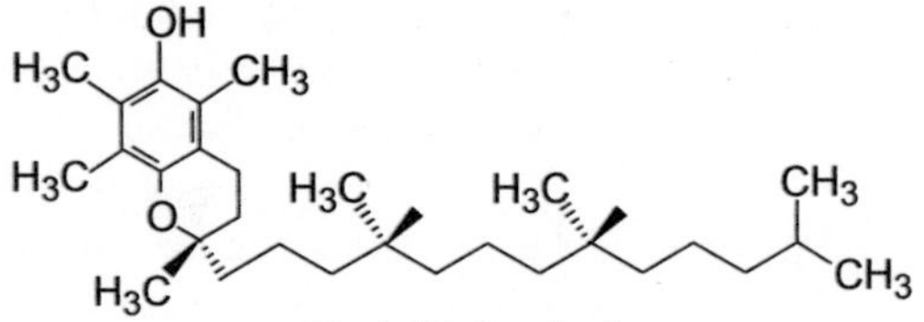

Vitamin E (α-tocopherol)

*Fig. 60 : **Vitamin E.***

Functions

- The principal role of vitamin E appears to be as an antioxidant. By accepting oxygen, vitamin E can prevent the oxidation of vitamin A in the intestinal tract, thus making vitamin A available for body use.
- Vitamin E also reduces the oxidation of the poly unsaturated fatty acids, thereby helping to maintain normal cell membrane.
- It protects the red blood cells against haemolysis.
- Vitamin E is required by animals, and presumably, by humans for the normal reproductive processes.
- It also plays an important protective role during ageing of cells.
- It is essesntial for the biosynthesis of coenzyme Q.
- It is essesntial for the normal functioning of muscles.
- It is essential for membrane structure and integrity.
- It is required for proper storage of creatine in skeletel muscles.

Higher intake of vitamin E may have a useful protective effect against the development of ischaemic heart disease. This is because high concentrations of vitamin E inhibit the oxidation of poly unsaturated fatty acids in plasma lipoproteins and this oxidation which is responsible for the initiation of atherosclerosis (deposition of fat in the coronory artery walls).

Sources - Wheat germ oil and corn germ oil are the rich natural sources. Vegetable oils and fats are good sources. Cereals and animal foods are fair sources of tocopherol.

Requirements - It is difficult to establish vitamin E requirements. The requirement depends mainly on the intake of poly unsaturated fatty acids. It is generally accepted that the intake of vitamin E should be 0.4 mg α-tocopherol equivalents / g dietary poly unsaturated fatty acid. This does not present any problem, since all foods which are rich sources of poly unsaturated fatty acids are also rich sources of vitamin E.

The levels that appear to be beneficial are of the order of 17-40 mg α-tocopherol/day, which is above what could be achieved by eating ordinary foods.

Adults : 25 - 30 mg / day

Absorption and storage -Vitamin E, like other fat soluble vitamins, is absorbed along with fat in the intestines. It is stored in the liver, muscle and body fat.

Deficiency

Vitamin E deficiency causes the following disorders in animals. They are Reproductive failure, Hemolysis of red blood cells, Muscular dystrophy, Hepatic necrosis and Exudative diasthesis. Higher dose administration of large doses of vitamin K produces haemolytic anaemia and Jaunice particularly in infants. The toxic effect is due to increased breakdown of RBC.

Vitamin K

It is a fat soluble vitamin. It was discovered by Dam in 1935. This vitamin is named as K due to it name Koagulation in German. It is a napthaquinone derivative.Vitamin K is known as the antihemorrhagic vitamin. It exists in three forms. They are Phylloquinone (vitamin K1), Menaquinones (vitamin K2) and Menadione (vitamin K3). Phylloquinone is found in green leafy vegetables. Menaquinones are synthesized by the intestinal bacteria. Menadione is a synthetic compound. Its metabolism leads to yield of phylloquinone.

Vitamin K are stable in heat.

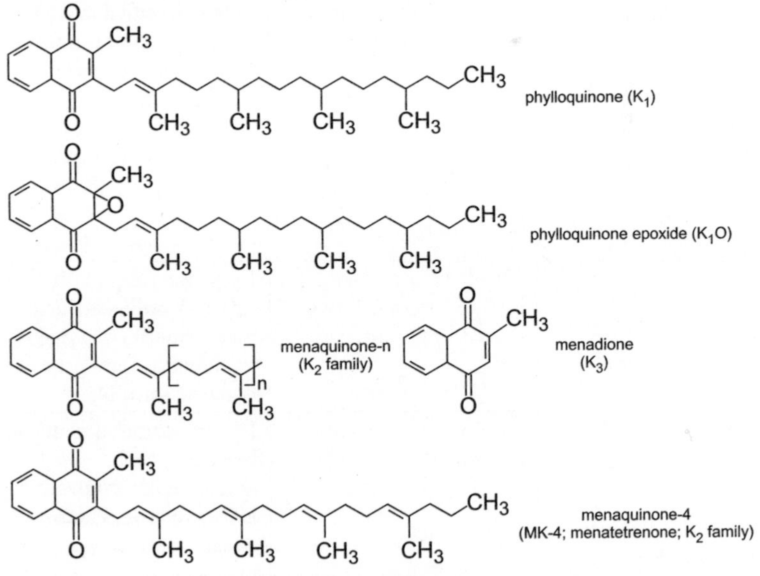

Fig. 61 : *Vitamin K.*

Functions

- Vitamin K is needed for the formation of prothrombin, a substance necessary for blood clotting.

Synthesis : Intestinal bacteria normally synthesize substantial amount of vitamin K. Vitamin K is transported along with LDL.

Sources : The best source of vitamin K are the green leafy vegetables eg. spinach, cabbage, kale etc. Good sources are cauliflower, wheat germ, etc. Carrots and potatos are fair sources.

Milk, meat and fish are poor sources.

Absorption and storage : Being fat-soluble, its absorption is enhanced by sufficient amount of bile salts mainly in the jejunum by the way of lymphatics. Liver stores appreciable amounts. It is present in blood stream in significant amount. All tissues contain small amounts of vitamin K.

Deficiency

The deficiency of vitamin K leads to a lowering of prothrombin level. This leads to increased clotting time of blood and leads to continuous bleeding from the damaged part and the victim die because of lose of blood. Vitamin K deficiency causes hemorrhagic disease of the newborn.

Fat-soluble vitamins

Nutrient	Function	Sources
Vitamin A (and its precursor*, beta-*carotene*) *A precursor is converted by the body.	Needed for *vision*, *healthy skin* and mucous membranes, bone and *tooth* growth, immune system health.	Vitamin A from animal sources (retinol): fortified milk, cheese, cream, butter, fortified margarine, eggs, *liver*. β-carotene (from plant sources): Leafy, dark green vegetables; dark orange fruits (apricots, cantaloupe) and vegetables (carrots, winter squash, sweet potatoes, pumpkin).
Vitamin D	Needed for proper absorption of *calcium*; stored in bones.	Egg yolks, *liver*, fatty fish, fortified milk, fortified margarine. When exposed to sunlight, the skin can make vitamin D.
Vitamin E	Antioxidant; protects cell walls	Polyunsaturated plant oils (soybean, corn, cottonseed, safflower); leafy green vegetables; wheat germ; whole-grain products; liver; egg yolks; nuts and seeds.

contd. next page

Nutrient	Function	Sources
Vitamin K	Needed for proper *blood* clotting	Leafy green vegetables and vegetables in the cabbage family; milk; also produced in *intestinal* tract by bacteria.

WATER SOLUBLE VITAMINS

The members of this group are B complex vitamins and vitamin C. They are readily soluble in water.

B Complex vitamins

Thiamine (B1)

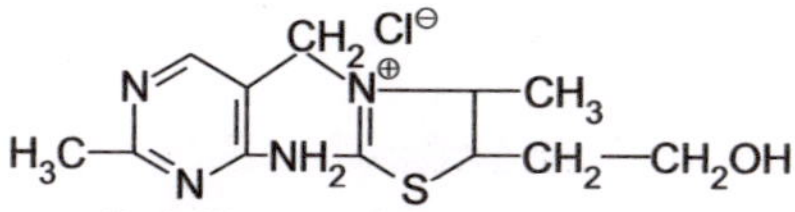

THIAMINE (Vitamin B_1)

*Fig. 62 : **Thiamine.***

It is a water soluble vitamin. It is isolated by Janson and Donath in 1936. It was synthesized by Williams in 1936 from cereals. Thiamine contains a pyrimidine ring and thiozole ring held by methylene bridge. This is a only natural compound with thiozole ring.

Functions : Thiamine act as a coenzyme in the form of thiamine pyrophosphate in many enzyme systems. These are involved in the breakdown of glucose to yield energy. Thiamine also aids in the formation of ribose. The adequate level of thiamine provides healthy nerves, a good mental outlook, a normal appetite and food digestion.

Sources : Meats, especially pork and liver are rich in thiamine. Dry beans, peanuts and egg are good sources.Whole grain breads and cereals supply about one third of the daily thiamine intake. Unpolished rice is a rich source of Thiamine.

Requirements : The requirement of thiamine depends on energy expenditure. RDA for vitamin B1 as follows. Infants: 0.3-0.5 mg / day; Children: 0.7-1.2 mg / day; Adults: 1.2-1.5 mg/day; Pregnant women and lactating women: 1.3-1.5 mg/day.

Absorption and storage : Free thiamine is readily absorbed from the small intestine. Excess thiamine administered is not stored in the tissues. A part of the excess thiamine is excreted in urine and same of it is destroyed by the enzyme thiaminase.

Deficiency : The symptoms of thiamine deficiency occur because the tissue cells are unable to receive sufficient energy from glucose. Therefore, they cannot carry out their normal functions. Early symptoms of thiamine deficiency include fatigue, irritability, depression and numbness of the leg and poor tone of the gastro intestinal tract together with constipation. The transketolase activity in RBC is decreased.

Carbohydrte metabolism is impaired.

Beriberi, sometimes called "rice-eaters disease" is another deficiency symptom which is often seen in people whose chief diet is refined rice and is the most severe form of thiamine deficiency. In adults two types of beri beri are indicated. They are wet beri beri and dry beri beri. Wet beri beri is a cardiovascular beri beri. It is characterized by edema of legs, face etc.

Dry beri beri is a neurological beri beri. Infantile beri beri is seed in infants born to mothers suffering from thiamine deficiency. Wernicke –korsakoff syndrome is a disorder. It is also known as cerebral beri beri. It is mostly seen in chronic alcoholics. The body demands of thimine increase in alcoholism.

Riboflavin (B2)

It is one of the water soluble vitamins. It is also called Vitamin B2. It takes part in an oxidation reduction reaction. It is isolated by Warburg and Christian in 1932. It is an orange yellow component. Structurally vitamin B2 consists of a ribitol moiety and a substituted isoalloxazine ring.

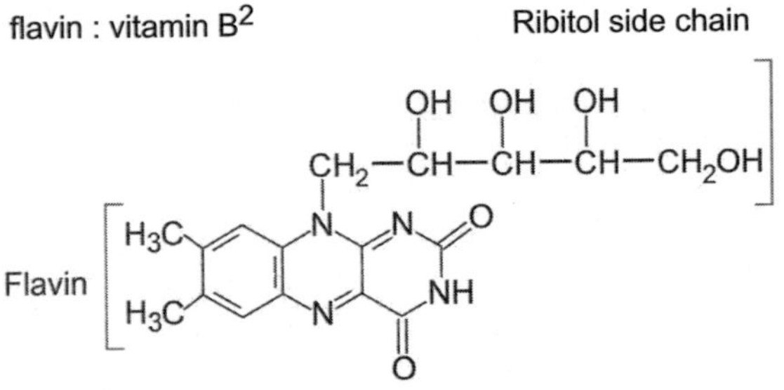

Fig. 63 : Riboflavin.

Functions

It is a component of FMN and FAD. They play a major role in various enzyme system. It is essential for metabolism in growth. It is a important component of acyl CoA dehydrogenase. Riboflavin is a constituent of a group of enzymes called "flavoproteins". As with thiamine, the enzymes are necessary in the break down of the glucose to form energy. Riboflavin is essential for a healthy skin and for good vision in bright light.

Sources : About half of the intake of riboflavin daily is furnished by milk alone and cheese is a good source, although some of the vitamin has been lost in the whey.

Requirements : Infants: 0.4-0.6 mg /day; Children: 0.8-1.2 mg/day; Adults male: 1.5-1.8 mg/day; Adults female: 1.1-1.4 mg/day; Pregnant women: 1.4 - 1.7 mg/day; Lactating women: 1.6 - 1.9 mg/day

Absorption and storage : The vitamin is phosphorylated in the intestinal mucosa during absorption. It is absorbed from the small intestine through the portal vein and is passed to all tissues being stored in the body. The major part is excreted in urine and a small part is metabolized in the body.

Deficiency : Riboflavin deficiency leads to cheilosis, a cracking of the skin at the corners of the lips and scaliness of the skin around the ears and nose. There may be redness and burning as well as itching of the eyes, and extreme sensitivity to strong light.

Niacin (B3)

It is a water soluble vitamin. It is a nicotinic acid. It is also known as pellagra preventive factor. It is a pyrimidine derivative. Structurally it is pyridine-3-carboxylic acid. It occurs in tissues as nicotinamide. Co enzyme of niacin are NAD and NADP. It is synthesized by essential amino acid tryptophan. Sixty milligram of tryptophan is equivalent to one milligram of niacin for the synthesis of niacin coenzymes.

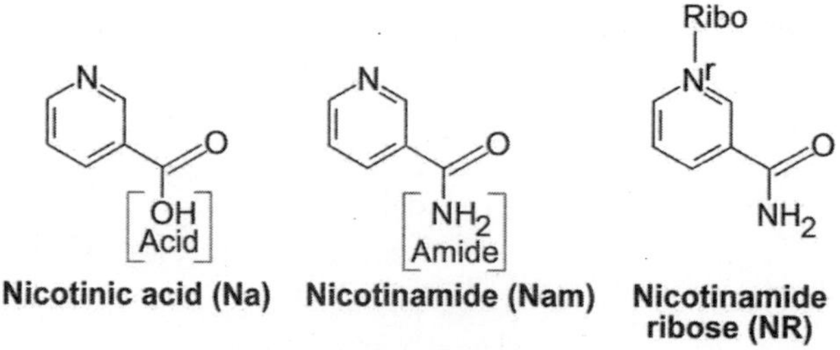

Nicotinic acid (Na)   Nicotinamide (Nam)   Nicotinamide ribose (NR)

Fig. 64 : Niacin.

Functions : Niacin is required for the stepwise breakdown of glucose to yield energy. Niacin is essential for the healthy skin, normal functions of the gastro intestial tract and maintenance of the nervous system. The coenzymes of niacin are involved in varieties of oxidation – reduction reactions.

Sources : The meat group especially organ meats and poultry, is the chief source of preformed niacin. Dark green leafy vegetables, whole grain, enriched breads and cereals are fair sources. Niacin is more stable to cooking procedures than thiamine or ascorbic acid (Vitamin C).

Requirments : Infants: 5-8 mg/day; Children: 9-16 mg / day; Adults male: 16-20 mg/day; Adults female: 12-16 mg/day; Pregnant women: 14 - 18 mg/day; Lactating women: 16 - 20 mg/day

Absorption and storage : Nicotinic acid and nicotinamide are absorbed from the intestine through the portal vein into the general circulation. Excess nicotinic acid is not stored in the body.

Deficiency-Pellagra (rough skin) is the major deficiency disease resulting from the lack of niacin. Dermatitis, especially of the skin exposed to the sun, soreness of the mouth, swelling of the tongue, diarrohea, and mental changes including depression, confusion, disorientation, and delirium are typical of the advancing stages of the disease, which ends to death if not treated. (The disease is sometimes referred to as the "4D'S" - dermatitis, diarrohea, dementia and death).

Vitamin B6 (Pyridoxine)

It was discovered by Gyorgy in 1934. Vitamin B6 collectively represents three compounds namely Pyridoxine, pyridoxal and pyridoxamine. Vitamin B6 is a pyridine derivative. The active form of Vitamin B6 is the co enzyme pyridoxal phosphate (PLP). It is synthesized from Pyridoxine, pyridoxal and pyridoxamine. Pyridoxine is 3-hydroxy 4,5-dihydroxy methyl-2-methyl pyridine. The metabollically active form of vitamin B6 is pyridoxal phosphate.

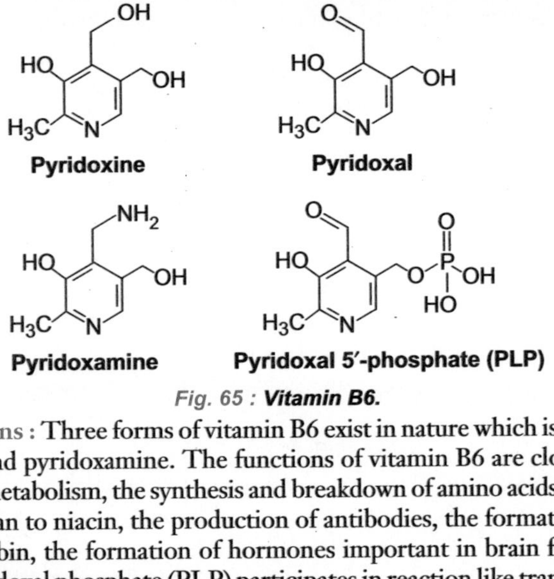

Fig. 65 : *Vitamin B6.*

Functions : Three forms of vitamin B6 exist in nature which is pyridoxine, pyridoxal and pyridoxamine. The functions of vitamin B6 are closely related to protein metabolism, the synthesis and breakdown of amino acids, conversion of tryptophan to niacin, the production of antibodies, the formation of heme in hemoglobin, the formation of hormones important in brain function and others. Pyridoxal phosphate (PLP) participates in reaction like transamination, decarboxylation, deamination, transsulfuration, condensation etc.,

Sources : Meat, especially organ meats, whole grain cereals, peanuts and wheat germ are rich sources. Milk and green vegetables supply smaller amounts. Egg yolk also rich source of vitamin B6.

Requirements : The average requirement is 13 g/g dietary protein. Infants: 0.3 mg/day; Children: 0.6-1.2 mg/day; Adults: 1.6-2 mg/day; Pregnant and lactating women: 2.5 mg/day

Absorption and storage : Pyridoxine is readily absorbed from the small intestine. The excess amount if ingested is not stored in the body but is excreted in urine.

Deficiency : Deficiency of vitamin B6 is extremely rare. Nervous disturbances such as irritability, insomnia, muscular weakness, fatigue and convulsion have been recorded in infants. The cause of the convulsions severe impairment of the activity of the enzyme glutamate decarboxylase, which is dependent on pyridoxal phosphate. The product of glutamate decarboxylase is GABA (γ-amino butyric acid) which is a regulatory neurotransmitter in the central nervous system.

Folic Acid (B9)

It was discovered by Day. It was isolated from spinach leaves. It is a water soluble vitamin. It is also called vitamin B9. Folic acid contains a pteridine group linked to para amino benzoic acid and l-glutamic acid.

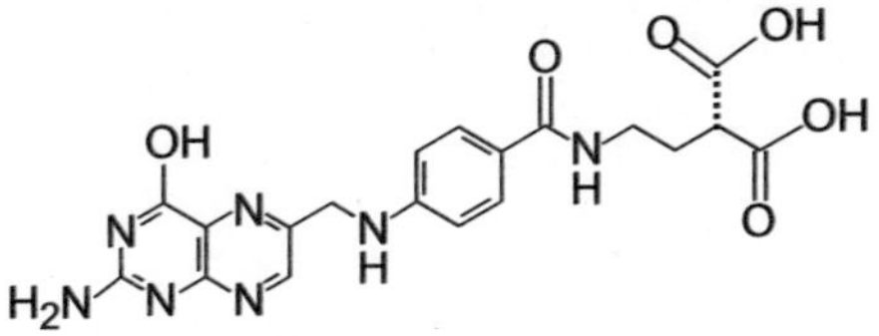

Fig. 66 : *Folic acid.*

It is slightly soluble in water and stable to heat. Tetra hydrofolate is a coenzyme of folic acid.

Functions : Folic acid serve as coenzymes in reactions involving the transfer of one carbon units like formyl and methyl groups. It participates in the reactions concerned with the synthesis of purine, pyrimidine and nucleic acids. It is essential for maturation of red blood cells. Folic acid is required for the metabolism of amino acids like histidine. Along with vitamin B12, folic acid helps in the trans methylation reactions. eg: uracil to thymine.

Sources : Folic acid is particularly present in green leafy vegetables, cauliflower and dried yeast. Egg, liver and kidneys are rich animal sources.

Requirements : Infants: 50 μg/day; Children: 100-300 μg/day; Adults: 400 μg/day; Pregnant women: 800 μg/day; Lactating women: 600 μg/day.

Absorption and storage : Absorption of folic acid takes place along the whole length of the mucosa of the small intestine. Folic acid about (5-15 mg/g) is in the liver and folate is also incorporated into the erythrocytes during erythropoiesis (Red blood cells production).

Deficiency : Deficiency of vitamin B12 also leads to functional folic acid deficiency. (i) Folic acid deficiency leads to megaloblastic anemia characterised by the release of large sized immature red blood cells into the circulation. (ii) Sprue and symptoms like glossitis and gastro intestinal disturbances have also been reported. iii. Macrocytic anemia of pregnancy responds to treatment with folic acid.

Vitamin B12

It is an anti-pernicious anaemia vitamin. Vitamin B12 is also called cobalamin. It is a water-soluble vitamin. It has a key role in the normal functioning of the brain and nervous system and the formation of red blood cells. It is a part of the vitamin B complex group. It is known to be the largest and most complex vitamin. It is essential for the formation of blood cells, nerve sheaths and various proteins, and is also necessary for growth.

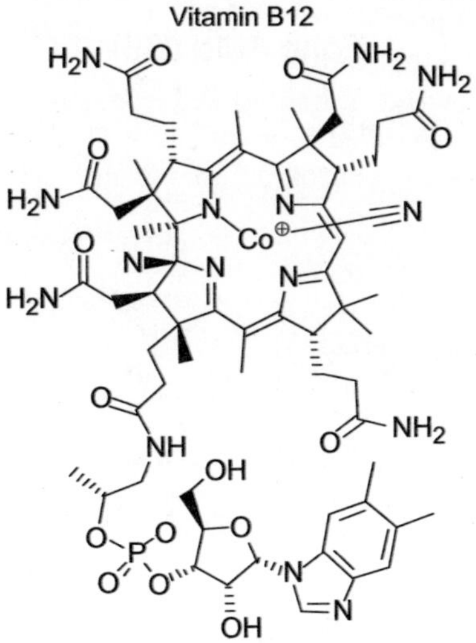

Fig. 67 : *Vitamin B12.*

Discovery and History : Vitamin B12 was discovered in 1926 by Georg Richard Minot and William Parry Murphy. In 1934, both scientists, as well as George Whipple, won a Nobel Prize for their work in treatment of pernicious anemia.

Structure of vitamin B12 : Structurally, vitamin B12 consists of a corrin nucleus attached with 5,6 dimethyl benzimidazole moitey, an aminopropanol unit, a ribose unit and a phosphate group. A cobalt atom (Co) is present at the centre of the corrin ring structure. One of the valencies of cobalt is filled by either CN - (cyano cobalamin) or H_2O (aqua cobalamin) or OH - (hydroxo cobalamin) or CH_3 (methyl cyano cobalamin).

Functions : Vitamin B12 is the most complex. The trace mineral cobalt is an essential part of the molecule. Vitamin B12 is required for the maturation of red blood cells in the bone marrow and for the synthesis of proteins. It is essential for the synthesis of nucleic acid. It prevents hyperglycemia. It synthesizes lipids from carbohydrates. It stimulates bone marrow to synthesis WBC and platelets. It prevents pernicious anemia

Sources : Milk, eggs, cheese, meet, fish and poultry supply ample amounts of vitamin B12.

Plant foods supply no vitamin B12 and use of an exclusively vegetarian diet for a long period of time will lead to symptoms of deficiency.

Requirements : Early estimates of vitamin B12 requirements were based on the amounts required to maintain normal RBC maturation in patients with pernicious anemia due to lack of intrinsic factor secretion. There is a considerable enterohepatic circulation of vitamin B12. It is secreted in the bile and re-absorbed in the small intestine. The average requirement of vitamin B12 is 3 µg/day. Infants: 0.3 µg / day; Children: 1-2 µg/day ; Adults: 3 µg/day; Pregnant and lactating women: 4 µg/day.

Absorption and Storage : For the absorption of vitamin B12 from the intestines, a factor called "Intrinsic Factor" (IF) secreted by the stomach is essential. Vitamin B12 is stored in fair amounts in the liver.

Deficiency : Pernicious anemia is the disease resulting from vitamin B12 deficiency. It is a genetic defect with the absence of intrinsic factor, hence the vitamin B12 in the diet cannot be absorbed. Vitamin B12 deficiency is usually due to insufficient diet and absorption problems.

It is rare in younger people and most commonly found in the elderly.

Symptoms include: Pernicious anemia (body cannot make enough red blood cells), Neuropathy (nerve damage), General fatigue, Loss of appetite and Gastric atrophy (chronic inflammation of the stomach).

Pantothenic Acid (B5)

It is also called chick anti dermatitis factor. The structure of pantothenic acid consists of an alanine chain in peptide linkage with dihydroxy, dimethyl butyric acid. Pantothenic acid is highly soluble in water.

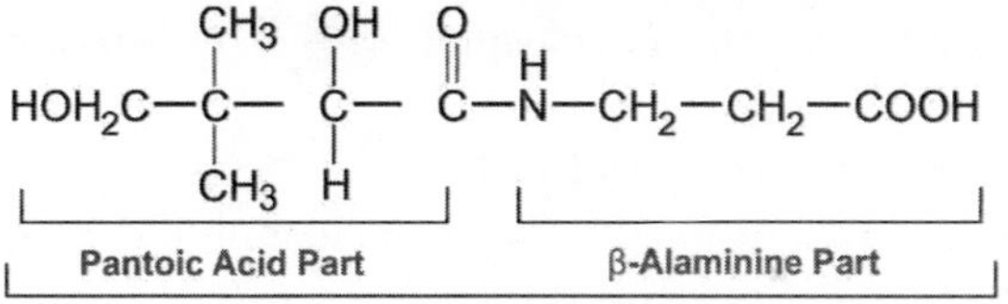

Pantothenic Acid

Fig. 68 : Pantothenic Acid.

Functions : Pantothenic acid exists in the free form and in combination with b-mercapto ethylamine, adenine ribose and phosphoric acid. The later form is called as co-enzyme A (CoA). The metabolic functions of pantothenic acid are due to its coenzyme derivative CoA, which participates in several metabolic reactions. CoA gains further importance after its conversion to form acetyl CoA. (i) Acetyl-CoA plays a key role in carbohydrate, protein and lipid metabolism. (ii) Acetyl-CoA is the precursor of cholesterol. It is the main source for the synthesis of cholesterol as well as steroid hormones. (iii) Acetyl-CoA combines with choline to form acetyl choline. iv. Some of the amino acids require CoA for their activation.

Sources : Dried yeast, liver, royal gelly are the rich sources of pantothenic acid. Egg yolk, meat, fish, milk are good sources.

Requirements : The recommended daily allowance of pantothenic acid as follows : Infants: 1.5-2.5 mg/day; Children: 5-8 mg/day; Adults: 5-12 mg/day; Pregnant and lactating women: 10-15 mg/day.

Absorption and storage : Pantothenic acid and its salts are readily absorbed from the small intestine through the portal vein into the general circulation. If ingested in excess of the requirements, it is not stored in the body; but is excreted in urine or metabolised by the tissues.

Deficiency : Deficiency of this vitamin results in nausea, vomitting, certain gastro intestinal tract disorders, inadequate growth, anemia, fatty liver and failure in gaining weights.

Biotin (B7)

It is a vitamin B7. It is a water soluble vitamin. It was discovered by Bateman in 1916. It is called **anti egg white injury factor**. It is also called coenzyme R because it is a growth factor for Rhizobium. Biotin is heterocyclic, sulphur containing monocarboxylic acid. The structure is formed by fusion of imidazole and thiophene rings with a valeric acid side chain. Biotin is sparingly soluble in cold water and is freely soluble in hot water.

Functions

1. Biotin is required as the co-factor for a synthesis of fatty acid from acetyl CoA through carboxylation reactions, it acts as the carrier for carbon dioxide.

$$CO_2 \text{ Acetyl CoA carboxylase}$$

eg. Acetyl CoA $\rightarrow\rightarrow\rightarrow\rightarrow\rightarrow\rightarrow\rightarrow\rightarrow\rightarrow\rightarrow\rightarrow\rightarrow\rightarrow\rightarrow\rightarrow\rightarrow$ Malonyl CoA.

2. It helps to maintain the skin and the nervous systems in good condition.

3. It assists in the deamination of amino acids like aspartic acid, serine and threonine.

4. It helps in the synthesis of purine.

5. Biotin is required for the the conversion of ornithine to citrulline. It is a major process of urea synthesis.

6. The conversion pyruvate to oxaloacetate by biotin dependent pyruvate carboxylase.

Source : Wheat germ, liver, peanut, and rice polishings are rich sources. Whole cereals, legumes, mutton and egg are good sources.

Requirements : Since, intestinal bacteria and diets supply biotin in adequate amounts the deficiency of this vitamin in human being is rare. Infants: 10-15 µg/day; Children: 20-40 µg/day; Adults: 50-60 µg/day.

Absorption and Storage : Biotin is readily absorbed from the small intestine through the portal vein into the general circulation. Excess of the requirements is not stored in the body but is mostly excreted in the urine.

Egg white injury factor (Avidin) : There is a protein in egg white called avidin which is responsible for producing egg white injury. Avidin binds with biotin tightly in the intestinal tract and prevents absorption of biotin from intestines. Avidin is denatured by cooking and then loses its ability to bind with biotin. The amount of avidin in uncooked egg white is relatively small, and problems of biotin deficiency have only occurred in people eating abnormally large amounts of raw eggs for many years.

Deficiency - Deficiency of biotin is rare in human beings. The symptoms of biotin deficiency include anaemia, loss of appetite, nausea, dermatitis. It also results in depression.

Vitamin C (Ascorbic Acid)

It is a water soluble versatile vitamin. It plays a major role in human health and disease. Vitamin C was discovered in 1912, isolated in 1928 by Szent-Gyorgyi, and first made in 1933 by Reichstein. Vitamin C is also called as ascorbic acid (**Fig.69**). It is a derivative of carbohydrate. The vast majority of animals and plants are able to synthesize vitamin C, through a sequence of enzyme-driven steps, which convert monosaccharides to vitamin C.

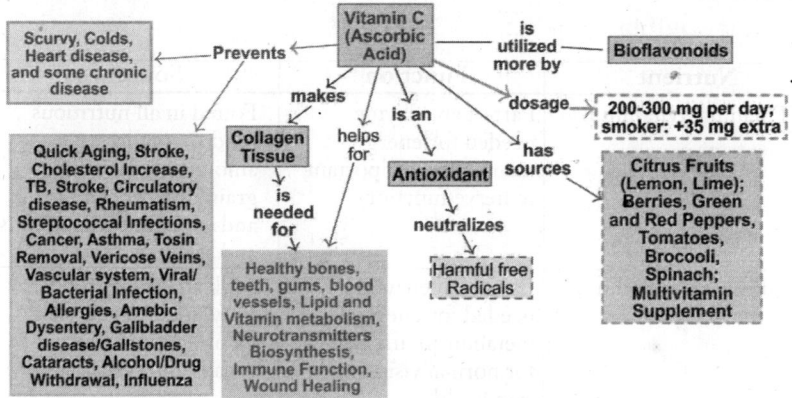

Fig. 69 : **Metabolism of Vitamin C.**

Functions : (**a**) Vitamin C is required for the biosynthesis of collagen, L-carnitine, and certain neurotransmitters; vitamin C is also involved in protein metabolism. This vitamin helps forming strong blood vessels, teeth firmly held in their sockets, and bones firmly held together.

(**b**) It has a general antioxidant role, especially in the regeneration of oxidized vitamin E in membranes.

(**c**) Ascorbic acid reduces the ferric ion (Fe^{3+}) to ferrous (Fe^{2+}) ion and thus helps in the absorption of iron. It is also essential for rapid healing of wounds.

Collagen is an essential component of connective tissue, which plays a vital role in wound healing.

Vitamin C is also an important physiological antioxidant and has been shown to regenerate other antioxidants within the body, including alpha-tocopherol (vitamin E).

Vitamin C plays an important role in immune function and improves the absorption of nonheme iron.

Sources : Raw fresh vegetables contain vitamin C, but some foods are more outstanding than others. Orange, grape, lime and lemon are especially rich in vitamin C.

Requirements : Recommended amount of vitamin C for different age group is as follows: Infants: 35 mg /day; Children: 40 mg / day; Adults: 45 mg/day; Pregnant women: 60 mg/day; Lactating women: 80 mg/day

Absorption and storage : Ascorbic acid is rapidly absorbed from the intestines and passed on through the portal vein to the general circulation. Liver and other organs and tissues have an optimal level of ascorbic acid. Excess intake does not increase further the optimal levels.

Deficiency : Severe deficiency of vitamin C leads to **scurvy**. This is characterised by easy bruising and hemorrhaging of the skin, lossening of the teeth, bleeding of the gums and distruption of the cartilages that support the skeleton.

Water-soluble vitamins

Nutrient	Function	Sources
Thiamine (vitamin B1)	Part of an enzyme needed for energy metabolism; important to nerve function	Found in all nutritious foods in moderate amounts: pork, whole-grain or enriched breads and cereals, legumes, nuts and seeds
Riboflavin (vitamin B2)	Part of an enzyme needed for energy metabolism; important for normal vision and skin health	Milk and milk products; leafy green vegetables; whole-grain, enriched breads and cereals
Niacin (vitamin B3)	Part of an enzyme needed for energy metabolism; important for nervous system, digestive system, and skin health	Meat, poultry, fish, whole-grain or enriched breads and cereals, vegetables (especially mushrooms, asparagus, and leafy green vegetables), peanut butter
Pantothenic acid	Part of an enzyme needed for energy metabolism	Widespread in foods
Biotin	Part of an enzyme needed for energy metabolism	Widespread in foods; also produced in intestinal tract by bacteria.
Pyridoxine (vitamin B6)	Part of an enzyme needed for protein metabolism; helps make red blood cells	Meat, fish, poultry, vegetables, fruits
Folic acid	Part of an enzyme needed for making DNA and new cells, especially red blood cells	Leafy green vegetables and legumes, seeds, orange juice, and liver; now added to most refined grains

contd. next page

Nutrient	Function	Sources
Cobalamin (vitamin B12)	Part of an enzyme needed for making new cells; important to nerve function found in plant foods	Meat, poultry, fish, seafood, eggs, milk and milk products; not
Ascorbic acid (vitamin C)	Antioxidant; part of an enzyme needed for protein metabolism; important for immune system health; aids in iron absorption	Found only in fruits and vegetables, especially citrus fruits, vegetables in the cabbage family, cantaloupe, strawberries, peppers, tomatoes, potatoes, lettuce, papayas, mangoes, kiwifruit

Introduction

Minerals are naturally occurring chemicals. They are stable at room temperature. They are represented by a chemical formula. They are formed usually abiogenic. They have ordered atomic arrangement. Mineral elements are inorganic substances. They are found in all body tissues and fluids. They are supplied by the diet as their salts. Eg. Sodium chloride. They may also be combined with organic compounds like iron in haemoglobin and sulphur in almost all proteins. The mineral elements are not destroyed in food preparations un like vitamins. However, they are soluble in water so that some loss will occur if cooking liquids are discarded.

Classification

Minerals are classified into two main groups. They are macro elements and micro elements. Macro elements are required in large amount (>100 mg/day) and present in large quantities in the body. Micro elements are required in small quantity (<100 mg/day) and also present in small amount in tissues and body fluids.

Macro elements or Bulk elements

Some of the macro elements are calcium, phosphorus, sodium, potassium, chlorine (as chloride), magnesium and sulphur.

CALCIUM

Calcium is a chemical element with symbol Ca and atomic number 20. Calcium is a soft grayish-yellow alkaline earth metal, fifth-most-abundant element by mass in the Earth's crust. Calcium is the most abundant mineral in the body. The total content of calcoim in an adult man is about 1-1.5kg.

Functions : Calcium is required for the development of skeletal system of our body. About 99% of the body calcium is found in the bones and teeth where it is combined with phosphorus and other elements to give rigidity to the bone. The bones also serve as the store house for calcium which is needed for a number of cellular functions. Cascade of blood clotting process required Calcium (Factor IV). Calcium triggers muscle contraction by interacting troponin C. It also activates ATPase, increases interaction between actin and myosin. It controls the transmission of nerve impulses. It influences the membrane structure and transport of water and several ions across it. It activates certain enzymes like lipase, adenosine triphosphatases, succinate dehydrogenase etc. Release of hormone from endocrine gland is facilitated by Ca^{2+}. It regulates microfilament and micro tubule mediated process such as endocytosis, exocytosis and cell motility. Calcium acts on myocardium and prolongs systole.

Distribution of calcium : Calcium is distributed in the body as follows: Serum: 9-11 mg/100 ml; C.S.F: 4.5-5 mg/100 ml; Muscle: 70 mg/100 gm; Nerve: 15 mg /100 gm

Calcium content of blood : The calcium content of plasma is fairly constant ranging from 9-11 mg/100 ml. This level is maintained constant in healthy individuals by the following factors: (i) the amount of calcium absorbed from food through the intestine and (ii) the secretion of parathyroid hormone which controls the level of calcium in blood.

Sources : Best sources of calcium are Milk and cheese. Other sources are Egg yolk, cabbage, beans and cauliflower.

Requirements : Recommended Dietary Allowances (RDAs) for Calcium are as follows. Infants: 200mg; 1-13 years: 700-1300mg; 14-18 years: 1300mg; Above 19 years: 100-1200mg/day.

Absorption and excretion : Calcium is actively absorbed in the duodenum. Calcium is taken in the diet as calcium salts of phosphate, carbonate, tartrate and oxalate. Calcium is excreted in the urine, bile and digestive secretions.

Factors affecting calcium absorption : Some of the factors which influence the absorption of the calcium in the intestine are: Vitamin D promotes absorption of calcium. High level of proteins in the diet helps to increase the absorption of calcium. Acidic environment in intestine favours calcium absorption. The beneficial effect of lactose is due to increased acidity (lactic acid) of the intestinal content which leads to increased calcium absorption. Excess of phosphate in diet lowers the calcium absorption. The ratio of calcium and phosphorus in the diet should be 1:1. Faulty absorption of fats leads to the presence of large amounts of fatty acids in the intestine were interferes with calcium absorption, as fatty acids form insoluble salt with calcium which are excreted in the feces. Oxalic acid present in certain food forms insoluble calcium oxalate which is excreted in the feces and interfere with the calcium absorption.

Disease status

Hypercalcemia : Elevation of calcium in serum are called hypercalcemia. It is associated with hyperparathyroidism. Symptoms include lethargy, muscle weakness, loss of appetite, constipation, nausea etc.

Hypocalcemia : It is characterized by fall in the serum calcium to below 7mg/dl, causing tetany. Symptoms include neuromuscular irritability and convulsions. Supplementation of oral calcium with vitamin D may prevent hypocalcemia.

Rickets : It is due to defective calcification of bone. It is a disease more directly related to vitamin D deficiency, but calcium and phophorus metabolism are also involved.

Osteoporosis : It is characterized by deminaeralization of bone resulting in the progressive loss of bone mass. High intake of calcium may reduce osteoporosis. Estrogen admininistration along with calcium and vitamin D to post menopausal women reduces the risk of fractures.

Osteopetrosis (marble bone disease) : It is characterized by increased bone density.

PHOSPHORUS

Phosphorus is a chemical element with symbol **P** and atomic number 15. As an element, phosphorus exists in two major forms—white phosphorus and red phosphorus—but because it is highly reactive, phosphorus is never found as a free element on Earth. Phosphorus is a mineral that makes up 1% of a person's total body weight. It is the second most abundant mineral in the body. It is present in every cell of the body. Most of the phosphorus in the body is found in the bones and teeth.

Functions : Phosphorus helps to filter out waste in kidneys. It also helps to manage energy use, growth, and repair of tissue and cells, produce DNA and RNA.

Blood content of phosphorus : The inorganic phosphorus content of blood in normal human adults ranges from 2.5 to 4.0 mg/dl and in children from 4 to 5 mg/dl.

Sources : This element is found in both animal and plant foods. Animal sources include fish, meat, egg, milk, liver and kidneys. Plant sources of phosphorus are nuts, beans, green vegetables and fruits.

Requirements : Infants: 0.24-0.4 g/day; Children: 0.8 -1.2 g/day; Adults: 0.8 g/day; During pregnancy and lactation: 1.5 g/day.

Absorption and excretion : Moderate amounts of fatty acid favour absorption of phosphorus. High calcium content in diet decreases the absorption of phosphorus. Phosphorus is excreted in the urine and feces.

Disease : A deficiency of phosphorus leads to **Rickets**. Low level of blood phosphorus characterised by defective bone and teeth formation. Excessively high levels of phosphorus in the blood is seen in hypoparathyrroidsm. High levels of phosphorus in blood only occur in people with severe kidney disease or severe dysfunction of their calcium regulation. High levels of phosphorus can affect your body's ability to effectively use other minerals, such as iron, calcium, magnesium, and zinc.

SODIUM

Sodium is a chemical element with symbol Na. In Latin it is called as natrium. Its atomic number is 11. It is a soft, silvery-white, highly reactive metal. Sodium is an alkali metal, being in group 1 of the periodic table, because it has a single electron in its outer shell that it readily donates, creating a positively charged atom—the Na^+ cation. Its only stable isotope is 23Na. it is a chief cation in the extracellular fluid. About 50% of sodium is found in bone, 40% in extracellular fluid and 10% in the soft tissue.

Functions : (i) It is the major component of the cations of extracellular fluid and exists in the body in association with the anions like chloride, bicarbonate, phosphate and lactate; (ii) Sodium ion is mainly associated with chloride and bicarbonate in the reguation of acidbase equilibrium. It maintains osmotic pressure of the body fluids and thus protects the body against excessive fluid loss; (iii) It plays an important role in the absorption of glucose and nutrients from small intestine; (iv) Sodium ions are involved in the maintenance of heart beat; (v) It maintains the normal neuromuscular functions and it functions in the permeability functioning of the cells.

Blood content of sodium : The normal level of sodium in blood is 310-340 mg/100 ml of blood. Red blood cells contain no sodium ions.

Sources : Sodium chloride (Common salt) is the main source of sodium. Bread, cheese and wheat germ are rich sources. Cauliflower, carrot and milk are also good sources of sodium.

Requirements : For adults the daily requirement is 5-10 grams/day.

Absorption and excretion : Sodium is completely absorbed from the gastro intestinal tract. About 95% of the sodium leaving the body is excreted in the urine since sodium is easily absorbed in the intestine.

Disease : Adrenal gland secretes hormones called adrenocortical steroids which regulates the metabolism of sodium. In the insufficiency of adrenocortical steroids the serum sodium level is decreased with an increase in sodium excretion. Hyponatremia (Addison's disease) the condition in which sodium level of blood is below normal and in hypernatremia (Cushing's syndrome) the sodium level of blood is above the normal range.

POTASSIUM

Potassium is a chemical element with symbol K and atomic number 19. It was first isolated from potash, the ashes of plants. In the periodic table, potassium is one of the alkali metals. It is a intracellular cation. It was given the name potassium because it was obtained from potash. It is an indispensible constituent of the body cells.

Functions : (i) Potassium is an important intracellular cation and needed for all cellular functions; (iii) It maintains the alkalinity of the bile and blood; (iii) It plays an important role in the regulation of acid-base balance in the cell; (iv) It influences cardiac muscle activity; (v) The glycolytic enzyme pyruvate kinase requires potassium for its maximal activity.

Sources : Most foods contain potassium. Animal sources include meat, fish , egg, milk, cheese. Vegetables like onion and carrot, fruits like banana, grapes and legumes such as beans also contain potassium.

Requirements : The normal intake of potassium in food is about 4 gm/day.

Absorption and excretion : Normally potassium is completely absorbed from the gastro intestinal tract and it is normally eliminated almost entirely in the urine and a small amount in the feces.

Disease : Deficiency of potassium leads to depression in cardiac and nervous system. Severe vomitting, diarrhoea, loss of appetite, fasting or starvation over a long period of time, may lead to deficiency of potassium. It also occurs during renal failure and shock. Fatigue, muscular weakness, heart and respiratory dysfunction are common signs of potassium deficiency.

Increase in the concentration of serum potassium is observed in renel failure.

CHLORINE

Chlorine is a chemical element with symbol Cl and atomic number 17. The second-lightest of the halogens, it appears between fluorine and bromine in the periodic table and its properties are mostly intermediate between them. It is a major component of sodium chloride (NaCl).

Functions : (i) Chloride ion is an essential for acid-base equilibrium; (ii) Chloride ion regulates the water balance and osmotic pressure; (iii) It is also important in the production of hydrochloric acid in the gastric juice. (iv) Chloride ion is important as an activator of amylase enzyme.

Sources : Sodium chloride is the main source for chlorine.

Requirements : The requirements of sodium chloride depend on the climate and occupation and on the salt content of the diet. Animal foods contain more NaCl than those of plant foods. Normal requirement of chloride is 5-10g /day as NaCl.

Absorption and excretion : Chloride is completely absorbed from the gastro intestinal tract. Chloride is mainly eliminated in the urine and also in the sweat.

Disease : Chloride deficiency occurs when loss of sodium is excessive in diarrhoea, sweating and some endocrine disturbances. This status is called hypochloremia.

Hyperchloremia is characterized by increased concentration of chloride ion due to dehydration, respiratory acidosis and Cushing's syndrome.

MAGNESIUM

Magnesium is a chemical element with symbol **Mg** and atomic number 12. It is a shiny gray solid which bears a close physical resemblance to the other five elements in the second column of periodic table. About 70 percent of the total magnesium content of the body is combined with calcium and phosphorus in the complex salts of bone. The reminder is in the soft tissues and body fluids. It is the principal cation of the soft tissues. Magnesium ions act as an activators for many of the phosphate group transferring enzymes and it also functions as a co-factor for oxidative phosphorylation.

Functions : Magnesium is a cofactor in more than 300 enzyme systems that regulate diverse biochemical reactions in the body, including protein synthesis, muscle and nerve function, blood glucose control, and blood pressure regulation. Magnesium is required for energy production, oxidative phosphorylation, and glycolysis. It contributes to the structural development of bone and is required for the synthesis of DNA, RNA, and the antioxidant glutathione. Magnesium also plays a role in the active transport of calcium and potassium ions across cell membranes, a process that is important to nerve impulse conduction, muscle contraction, and normal heart rhythm.

Blood content of magnesium : The normal level of magnesium in blood is 1-3 mg/100 ml.

Sources : Milk, egg, cauliflower, cabbage and fruits are rich sources of magnesium.

Requirements : Infants: 100-150 mg/day; Children: 150-200 mg/day; Adults: 200-300 mg/day

Absorption and excretion : Like calcium the salts of magnesium are rather insoluble and greater part of the daily ingested magnesium is not absorbed. Parathyroid hormone increases the magnesium absorption. Very high intake of fat, phosphate and calcium decrease the absorption of magnesium. The major quantity of magnesium is excreted in the feces and remaining is excreted through urine.

Disease : Magnesium deficiency causes depression, muscular weakness, and liability to convulsions. Severe magnesium deficiency can result in hypocalcemia or hypokalemia (low serum calcium or potassium levels, respectively) because mineral homeostasis is disrupted. Magnesium is involved in bone formation and influences the activities of osteoblasts and osteoclasts. Magnesium also affects the concentrations of both parathyroid hormone and the active form of vitamin D, which are major regulators of bone homeostasis.

Magnesium deficiency is related to factors that promote headaches, including neurotransmitter release and vasoconstriction.

SULPHUR

Sulfur is the third most abundant mineral in the body, about half concentrated in your muscles, skin and bones, and is essential for life. Sulfur or sulphur is a chemical element with symbol S and atomic number 16. It is abundant, multivalent, and nonmetallic.sulphur. It is available in our body in an organic form. Sulphur is present in the amino acids such as methionine, cystine and cysteine. Thus, it is present in all proteins in the body. Connective tissue, skin, hair and nails are rich in sulphur. Thiamine and biotin also contain sulphur.

Functions : Sulfur makes up vital amino acids used to create protein for cells and tissues and for hormones, enzymes, and antibodies. Sulfur is needed for insulin production. Sulfur detoxifies at the cellular level and relieves pain. Sulfur builds flexible cells in the arteries and veins. Sulfur has been called nature's "beauty mineral" because it keeps your complexion clear and youthful and hair glossy and smooth. Collagen production in your body depends on sulfur to create healthy skin and heal scars. Methylsulfonylmethane (MSM) has an amazing anti-parasitic action. MSM has anti-allergic properties. MSM has an ability to bind to mucuous membranes and form a natural block against allergens.

Sources : Sulphur intake is mainly in the form of cystine and methionine present in protein. Other compounds present in the food contribute small amounts of sulphur.

Requirements : A diet that is adequate in protein supplies required amounts of sulphur.

Absorption and excretion : Inorganic sulphate (SO_4^{2-}) is absorbed as such from the intestine into the portal circulation. Sulphur is excreted in the urine.

Deficiency : No specific sulphur deficiency has been observed in human beings.

IRON

Iron is a mineral found in every cell in the body. It is vital for both physical health and mental well-being. There are two types of iron in food. They are Heme iron, derived from the haemoglobin and myoglobin found in meat tissue. Non-heme iron, derived mainly from cereals, legumes, fruit and vegetables. The total content of iron in the adult body is only 3 to 5 grams.

Functions

Iron has three main functions :

1. Carrying oxygen from the lungs to the rest of the body- Most of the iron is present in haemoglobin. Haemoglobin is carried in the circulation by the red blood cells. It picks up oxygen in the lungs and transports the oxygen to the tissues so that oxidation reactions can take place in the cells. From the cells the haemoglobin carries CO_2 to the lungs to be exchanged.

2. Maintaining a healthy immune system - Body protects itself from antigens, which includes virus, bacteria and foreign substances that make sick.

3. Aiding energy production -Iron is constituent of several enzymes including : iron catalase, peroxidase, and cytochrome enzymes.

Sources : The richest animal sources of iron are beef, liver, heart, kidneys, spleen, egg yolk and land snail. Plant sources of iron include whole wheat and its products, green vegetables, onions, coconuts and fresh fruits.

Requirements : Iron requriements are influenced by the availability of iron present in foods. Iron present in cereals, legumes and green leafy vegetables is available to a lesser extent than that present in egg, meat and fish. In view of this, iron requriements of persons consuming a predominantly cereal based diet, will be greater than those consuming large quantities of meat and egg. About 10 percent of the ingested iron is only absorbed.

Infants: 10-15 mg/day; Children: 15 mg/day; Adults: 18 mg/day; During pregnancy and lactation: 40 mg/day.

Absorption and excretion : Under normal conditions, very little dietary iron is absorbed which take place mainly in the stomach and the duodenum. Infants and children absorb a higher percentage of iron from food than adults. Iron deficiency in infants is due to a dietary deficiency. Iron deficient children absorb twice as much as that of normal children. The body stores of iron are conserved very efficiently. Only lesser amounts are excreted in the urine, feces and sweat.

Deficiency : Insufficient iron in the diet causes iron deficiency **anemia**. This is common in growing children and pregnant women. Insufficiency may be brought about by blood loss during menstruation, inadequate intake of iron and hookworm infestation.

COPPER

Copper is incorporated into a variety of proteins and metalloenzymes which perform essential metabolic **functions**. This micronutrient is necessary for the proper growth, development, and maintenance of bone, connective tissue, brain, heart and many other body organs. The importance of copper for the formation of haemoglobin was studied by Hart and co-workers in 1928. Later studies indicated that copper has many functions.

Functions : (i) Copper is essential for the synthesis of haemoglobin. (ii) It is needed for the synthesis of collagen, melanin and phopholipids. (iii) It is a constituent of several enzymes. (iv) Three copper containing proteins namely cerebrocuprein, erythrocuprein and hepatocuprein are present in brain, RBC and liver respectively.

Sources : Copper is present in minute quantities in most foods, including liver, kidneys, shell fish and meat. Plant sources of copper are nuts and legumes.

Requirements : Daily requirement of copper is 0.05-0.85 mg/kg body weight for children and 2 mg/day for adults.

Absorption and excretion : Absorption of copper into the blood stream occurs via the villi of the small intestine. About 30 percent of the dietary copper is absorbed in the duodenum. Only 10-60 µg of copper is excreted in normal urine in 24 hours.

Disease : Normally people have enough copper in the foods they eat. Menkes disease (kinky hair syndrome) is a very rare disorder of copper metabolism that is present before birth. It occurs in male infants. Lack of copper may lead to anemia and osteoporosis. In large amounts, copper is poisonous. A rare inherited disorder, Wilson's disease, causes deposits of copper in the liver, brain, and other organs. The increased copper in these tissues leads to hepatitis, kidney problems, brain disorders, and other problems.

IODINE

Iodine is a chemical element with symbol I and atomic number 53. The heaviest of the stable halogens, it exists as a lustrous, purple-black metallic solid at standard conditions that sublimes readily to form a violet gas. **Iodine** is needed for the normal metabolism of cells. **Humans** need **iodine** for normal thyroid **function,** and for the production of thyroid hormones.the total body contains about 20mg of iodine. Most of the iodine (80%) is available in thyroid gland.

Functions :

Metabolic Rate : Iodine influences the functioning of thyroid glands by assisting in the production of hormones, which are directly responsible for controlling the body's base metabolic rate. Certain hormones, like thyroxin and triodothyronine, influence heart rate, blood pressure, body weight and temperature.

Energy Levels : Iodine also plays an important role in maintaining optimal energy levels of the body by ensuring the efficient utilization of calories, without allowing them to be deposited as excess fats.

Healthy nails, hair and teeth : The health benefits of iodine include the formation of healthy and shiny skin, teeth and hair.

Reproductive System : It helps in the normal growth and maturity of reproductive organs. A sufficient quantity of iodine in pregnant women is essential to prevent stillbirths or neurocognitive conditions like cretinism in the newborn babies.

Immune System Strength : Iodine is itself a scavenger of free hydroxyl radicals, and that, like vitamin-C, it also stimulates and increases the activity of antioxidants throughout the body to provide a strong defensive measure against various diseases, including heart disease and cancer.

Fibrocystic disease : Iodine can significantly reduce conditions like fibrosis, turgidity, and breast tenderness. Iodine acts as a relief for fibrocystic diseases and is widely used in therapies, both alternative and modern.

Sources : Vegetables grown in iodine-rich soil will naturally be the good sources of iodine. The animal sources of iodine include milk, sea fish, shell fish and crabs.

Requirements : This is normally supplied by an ordinary well balanced diet and by drinking water except in mountainous regions. Iodised salt may be added when cooking foods in areas with insufficient natural sources of iodine. Infants: 50-100 g/day; Adults: 100-150 g/day; Pregnant women: 200 g/day.

Absorption and excretion : Absorption is through the villi of the small intestine into the blood stream and 90 percent of the iodine of the thyroid gland is in organic combination and stored in the follicular colloids as thyroglobulin. Inorganic iodine is mostly excreted by the kidneys, liver, skin, lungs and intestine and in milk. About 10 percent of circulating organic iodine is excreted in feces.

Deficiency : A deficiency of iodine leads to a decreased production of thyroxine, and inturn a lowered rate of energy metabolism. In an attempt to produce more thyriod hormones the thyroid gland enlarges. This condition is called **simple or endemic goiter**. In a mild deficiency the only symptom noted is slight enlargement of the thyroid gland visible at the neckline. However, if the condition persists, the women who has a simple goiter and who fails to get sufficient iodine during pregnancy will be unable to supply the fetus with its needs: thus, the baby is more severely affected than its mother. People born in areas which lack of iodine in the water and soil have a tendency to develop 'goitre', a condition characterised by subnormal metabolic activities. The condition may be reversed if treated sufficiently with iodine at the early stage itself.

FLUORINE

Fluorine is a chemical element with symbol F and atomic number 9. It is the lightest halogen and exists as a highly toxic pale yellow diatomic gas at standard conditions. *It is mostly found in bones and teeth.*

Functions : Fluorine exists in the body in compounds called fluorides. (i) Lesser amount of fluoride enter into the complex calcium salts that form tooth enamel; (ii) Fluorides may also be useful in maintaining bone structure. It is necessary for the prevention of dental caries; (iii) Fluoride ions inhibit the metabolism of oral bacterial enzymes and diminish the local production of acids which are important in the production of dental caries; (iv) It is in combination with vitamin D, required for the treatment of osteoporosis.

Sources : The chief source of fluorine is in the form of fluoride in drinking water.

Requiremnts : Fluorine is found is small amounts in normal bones and teeth. Since water containing 1-2 ppm (Parts per million) prevents dental caries and does not do any harm, the fluorine requirements of the body are met by the quantity normally present in drinking water (1-2 ppm) in most of the regions.

Absorption and excretion : Absorption of fluoride is via the small intestine into the blood stream. Fluoride is excreted in the urine and in the sweat, and by intestinal mucosa. Most of the fluorides that are not retained by the bones and teeth is excreted rapidly into the urine.

Deficiency : The absence of fluorine in the diet causes dental caries.

Toxicity : Excess of fluorine (above 5 ppm) causes chalky white patches on the teeth. If this is not treated in time, the patches change to a brown colour which later develop into holes.

ZINC

Zinc is a chemical element with the symbol Zn and atomic number 30. It is the first element in group 12 of the periodic table. About 2 to 3 grams of zinc are found in the body. Like iron, zinc is absorbed according to the body needs.

Functions : It is required for normal growth and sexual maturation. It helps to transfers carbon dioxide from tissues to the lungs. It increases the production of insulin by the pancreas. It improves the synthesis of proteins. It maintain normal sensitivity to taste.

Sources : Zinc is widely distributed in animal and plant foods that are good sources of protein. Meats, egg, liver, sea food, legumes, nuts, milk wholegrain and cereals are good sources. People who eat a normal diet adequate in protein are not likely to develop zinc deficiency.

Requirements : Daily requirements of zinc are given as follows. Infants: 3-5 mg/day; Children: 10-15 mg/day; Adults: 15 mg/day; During pregnancy and lactation: 20-25 mg/day.

Absorption and excretion : Zinc present in animal foods are well absorbed in the small intestine, especially from the duodenum. Zinc present in plant foods are poorly absorbed due to the presence of phytic acid which interferes with its absorption. Zinc is mostly excreted in the feces.

Deficiency : The diet that has low zinc level leads to dwarfism and retarded sexual development. Zinc deficiency leads to diminished sensitivity to taste (hypogeusia) and to a decrease in odour sensitivity (hyposmia).

COBALT

Cobalt is a chemical element with symbol Co and atomic number 27. Cobalt is found in the Earth's crust only in chemically combined form.

Functions : Cobalt occurs in small amount in all tissues, higher concentrations occuring in liver and kidneys. Most of the cobalt is present in vitamin B_{12} which is necessary for red blood cells maturation.

Sources : It is largely available in food.

Requirements : Cobalt deficiency has not been observed in human beings. Cobalt requirements, if any, appear to be met by traces of cobalt found in foods. Human beings require a dietary source of vitamin B12 which is not synthesized by the body.

Absorption and excretion : Cobalt is readily absorbed from the small intestine. About 65% of ingested cobalt is excreted in the urine and the remainder in the feces.

Deficiency : Cobalt deficiency is rare in human beings.

MANGANESE

Manganese is a chemical element with symbol Mn and atomic number 25. It is not found as a free element in nature. It is often found in minerals in combination with iron.

Functions : (i) Manganese is essential for normal bone structure, reproduction and the normal functioning of the central nervous system; (ii) Manganese activates isocitrate dehydrogenase and phosphotransferases; (iii) Pyruvate carboxylase and superoxide dismutase contain tightly bound manganese; (iv) Manganese ions activate glycosyl transferases which is concerned with synthesis of muco polysaccharides of cartilages and also associated with the synthesis of glycoproteins; (v) Arginase an enzyme which is involved in the urea cycle is activated by manganese ions.

Sources : Nuts and whole grains are rich sources and vegetables and fruits are good sources of manganese.

Requirements : The average dietary intake of 2.5 to 9.0 mg/day is sufficient to meet the daily needs.

Absorption and excretion : Manganese is readily absorbed in the small intestine. Normally 3 to 4 percent of manganese present in the diet is absorbed. Large quantity of manganese is excreted mostly in the feces. Only very small quantities of manganese is excreted in the urine.

Deficiency : The deficiency of manganese leads to impaired growth and skeletal abnormalities.

CHROMIUM

Chromium is a chemical element with symbol Cr and atomic number 24. It is the first element in group 6. It is a steely-grey, lustrous, hard and brittle metal which takes a high polish, resists tarnishing, and has a high melting point. Chromium occurs in trace amounts in human and animal tissues.

Functions : Chromium plays an important role in carbohydrate, lipid and protein metabolism. It potentiates the action of insulin in accelerating the utilization of glucose.

Sources : The best source of chromium is brewer's yeast, but many people do not use brewer's yeast because it causes bloating (abdominal distention) and nausea. Other good sources of chromium include Beef, Liver, Eggs, Chicken, Oysters, Wheat germ, Green peppers, Apples, Bananas, Spinach. Black pepper, butter, and molasses are also good sources of chromium.

Requirements : The exact requirements are not known. Average diets can meet the chromium requirements.

Absorption and excretion : It is readily absorbed in the small intestine. It is mobilized from the tissues in response to glucose administration. Chromium is mainly excreted in urine, a small amount in bile and feces.

Deficiency : Chromium deficiency may be seen as impaired glucose tolerance. It is seen in older people with type 2 diabetes and in infants with protein-calorie malnutrition. Supplementation of chromium helps with management of these conditions, but it is not a substitute for other treatment. Because of the low absorption and high excretion rates of chromium, toxicity is not common.

MOLYBDENUM

Molybdenum is a chemical element with symbol **Mo** and atomic number 42. The element was discovered by Carl Wilhelm Scheele in 1778. The metal was first isolated in 1781 by Peter Jacob Hjelm. Molybdenum also occurs in traces in human and animal tissues. Molybdenum is classified as a metallic element and found widely in nature in nitrogen-fixing bacteria. Molybdenum is essential in trace amounts for human, animal and plant health.

In humans and animals, molybdenum serves mainly as an essential cofactor of enzymes and aids in the metabolism of fats and carbohydrates. Humans need only very small amounts of molybdenum, which are easily attained through a healthy diet. Molybdenum deficiency is very rare in humans, so molybdenum supplements are rarely needed.

Functions : Molybdenum is an essential component of the xanthine oxidase. It is also present in nitrate reductase in plants, and nitrogenase, which functions in nitrogen fixation by micro organisms. Trace amount of molybdenum are requried for the maintenance of normal levels of xanthine oxidase in animal tissues.

Source : It is available in average diets. Human milk also contains molybdenum.

Requirements : Adequate amounts of molybdenum are present in average diets. The Recommended Dietary Allowance (RDA) of molybdenum for men and women is 45 µg/day. Infant from birth to six months is 2 µg/day and for infants seven to 12 months is 3 µg/day. The RDA for children from ages one to three is 17 µg, children from four to eight is 22 µg/day, children nine to 13 is 34 µg/day and children 14 to 18 years old is 43 µg/day.

Absorption and excretion : About 50 to 70 percent of the ingested molybdenum is readily absorbed in the small intestine. About half of the absorbed molybdenum is excreted in urine.

Deficiency : Molybdenum deficiency is rare in human beings. A congenital molybdenum cofactor deficiency disease, seen in infants. It is an inability to synthesize molybdenum cofactor, a heterocyclic molecule that binds molybdenum at the active site in all known human enzymes that use molybdenum. The resulting deficiency results in high levels of sulfite and urate, and neurological damage

SELENIUM

It was discovered by Berzelius in 1817, as an impurity in sulphuric acid. It occurs in various minerals, together with **sulphur.** Selenium is present in foods of plant origin grown in selenium rich soils. Selenium is a nutritionally essential element. People need selenium for healthy joints, heart and eyes. It plays a critical role in DNA synthesis, the immune system and the reproductive system. It also helps fight cancer and other diseases. It is an element with symbol Se and atomic number 34.

Functions : Selenium is essential for normal growth, fertility and for the prevention of a wide variety of diseases in animals. Selenium is a constituent of the enzyme Glutathione peroxidase (GPx), a selenoprotein. This enzyme is the protective agent against the accumulation of H_2O_2 (Hydrogen peroxide), and organic peroxides within cells. It is involved in immune mechanisms, ubiquinone synthesis and mitochondrial ATP biosynthesis. Selenium has close metabolic relationship with vitamin E for curing certain diseases.

Sources : Selenium is largely present in different foods. The variation depends on the selenium content of the soil. Selenium is found in many different foods. Seafood and organ meats are the richest sources. Other foods that have selenium include muscle meats, grains and dairy products. It can also be found in drinking water in some places.

Requirements : Any normal diet can meet the daily requirement of selenium.

Deficiency : Selenium deficiency is very rarely seen in human beings. However, necrosis and muscular dystrophy are associated with selenium deficiency.

Introduction

Peoples believe that blood has magical qualities. Blood helps to maintain homeostasis. Blood is an essence of life because uncontrolled lose of it can results in death. Blood is a constantly circulating fluid providing the body with nutrition, oxygen and waste removal. Blood is mostly liquid, with numerous cells and proteins suspended in it, making blood "thicker" than pure water. The blood is a fluid connective tissue. It consists of liquid plasma and cells. The plasma makes up 55% of the total volume and 45% of cells or formed elements. The total blood volume in human female is about 4-5 litres and 5-6 litres in males. The blood plasma is a slightly alkaline fluid. It is straw coloured. Blood is a fluid which circulates in a closed system of blood vessels, in multicellular and highly complex vertebrate animals.

Formation of blood / Origin of Blood

Biosynthesis of blood is known as haematopoiesis. The process consist of production, development and maturation of cellular elements of blood. In adults the production of blood cells is carried out by the bone marrow while in the developing fetus haematopoisis takes place in extra medullary regions till the bone marrow is matured and functional. All cellular blood components are derived from haematopoietic stem cells. In a healthy adult person, approximately 10^{11}–10^{12} new blood cells are produced daily in order to maintain steady state levels in the peripheral circulation. Haematopoietic stem cells (HSCs) reside in the medulla of the bone (bone marrow) and have the unique ability to give rise to all of the different mature blood cell types and tissues. All blood cells formation process are divided into three lineages. they are lymphoid lineage, myeloid lineage and erythroid lineage.

Erythroid lineage synthesizes RBC and platelets. T-Cells, B-Cells and NK cells are synthesized through lymphoid lineage and Neutrophils, Basophil, eosinophil, Monocytes, macrophages, Mast cells are synthesized through myeloid lineage.

In developing embryos, blood formation occurs in aggregates of blood cells in the yolk sac, called blood islands. As development progresses, blood formation occurs in the spleen, liver and lymph nodes. When bone marrow develops, it eventually assumes the task of forming most of the blood cells for the entire organism. However, maturation, activation, and some proliferation of lymphoid cells occurs in the spleen, thymus, and lymph nodes. In children, haematopoiesis occurs in the marrow of the long bones such as the femur and tibia. In adults, it occurs mainly in the pelvis, cranium, vertebrae, and sternum.

Composition

Blood is a type of connective tissue. Blood contains 55% of the plasma and 45% of the cellular fraction. The water constitutes about 80% of blood by weight. The cellular fractions composed of erythrocytes, leucocytes and thrombocytes.

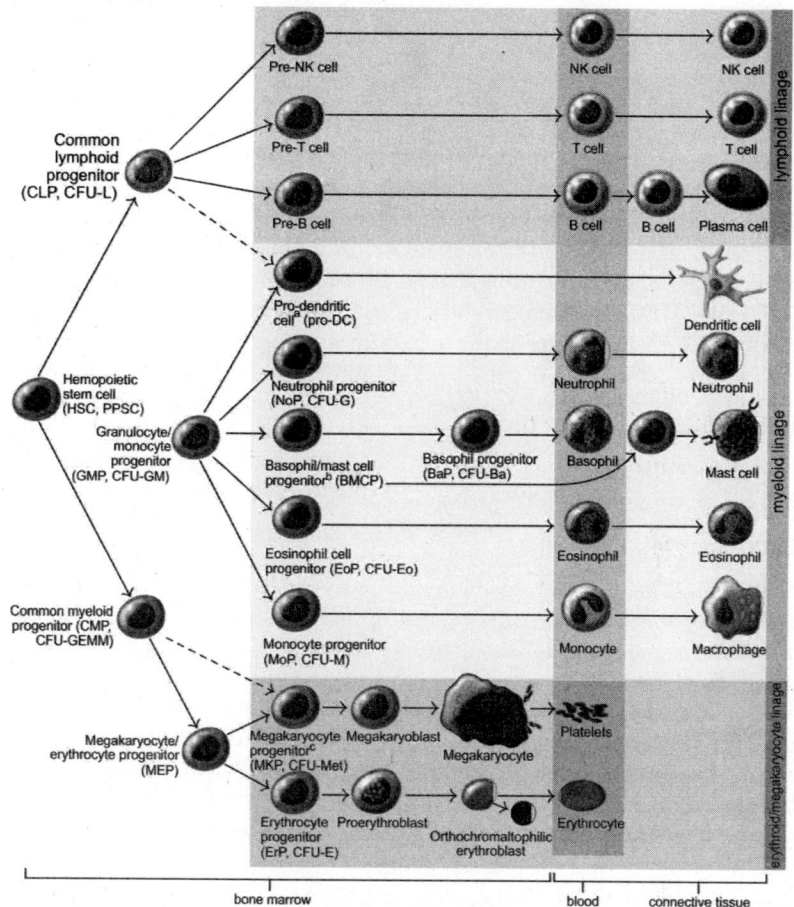

Fig. 70 : **Haemopoiesis.**

Blood Plasma

It is a clear, straw-coloured, liquid portion of the blood in which the cells and platelets are suspended. About 55% of blood is blood plasma. Human contains 2.7–3 liters of plasma. It is essentially an aqueous solution containing 92% water, 8% blood plasma proteins, and trace amounts of other materials. Plasma circulates dissolved nutrients, such as glucose, amino acids and fatty acids (dissolved in the blood or bound to plasma proteins), and removes waste products, such as carbon dioxide, urea and lactic acid. The term **serum** refers to plasma from which the clotting proteins have been removed. Most of the proteins remaining are albumin and immunoglobulins. Plasma contains 91% water, 7% protein and 2% other components such as ions, nutrients.

Plasma proteins

About 58% plasma proteins are albumins. They are produced by the liver, acts as a carrier molecule, blood buffer, and used in the regulation of osmotic pressure. About 4% of total plasma proteins are Fibrinogen. It involves in clotting mechanisms. Globulins accounts for 38% of total plasma proteins. They are used as transport proteins (egs alpha and beta) and circulate as antibodies (gamma). Other proteins available as metabolic enzymes, antibacterial proteins, and hormones.

Physical Characteristics of Blood

Blood makes up 8% of total body weight. Blood is more dense than water and five times more viscous. Blood is slightly alkaline in nature (pH 7.35-7.45). Blood temperature is Slightly higher than body temperature (100°F). The total blood volume in average adult is 5 liters (Male = 5-6 liters, female 4-5 liters). **Serum** is plasma without clotting factors.

Formed Elements of blood

Blood contains 45% cellular fractions. Among the cellular fraction, RBC accounts for about 95%. Remaining 5% made up of WBC, Platelets.

Erythrocytes (RBCs)

In blood, RBCs are available 700 times higher than WBC. It is a Small biconcave, anucleate disks like structure. RBCs are thin in center and thick around rim. Bioconcave structure increases surface area thereby maximizing diffusion of gases. It is easier to gas to enter and exit. One third of the RBC area is occupied with oxygen-binding protein, **haemoglobin.** Haemoglobin has a quaternary structure. It contains heme, each eume contains one iron atom. Iron is responsible for oxygen bining. Forms of haemoglobin are oxy haemoglobin, deoxy haemoglobin and carbamino haemoglobin. RBCs contain a protein (**spectrin**) which forms a net and is deformable, enabling RBCs to change shapes. RBC count in (per mm cubed) males is 4.8 - 6.2 million, female is 4.2 - 5.4 million. RBC production (erythropoiesis) occurs in three stages: 1) production of ribosomes, 2) synthesis of haemoglobin, and 3) ejection of erythrocyte's nucleus and organelles. RBC life span is 100-120 days. RBCs may become trapped and fragment in the smaller circulatory channels (capillaries) and are then phagocytized by macrophages. RBCs are broken down into globin and heme which is further broken down into iron and the greenish pigment biliverdin (then converted into bilirubin). Raw materials (iron & B vitamins absorbed by intestine) and erythropoeitin promote eryhtropoeisis in marrow. Erythrocytes enter blood stream and circulate for 120 days. Aged/damaged RBCs engulfed by macrophages of liver and spleen, haemoglobin sequestered by liver.

Structure of Haemoglobin-Refer Page 49.

Production of RBC

In response to decreased blood oxygen, the kidneys release erythropoietin into the circulation. The increased erythropoietin stimulates red marrow to produce more RBCs. This process increases blood oxygen levels.

Haemoglobin breakdown

In macrophages, the globin part of haemoglobin is broken down into aminoacids and used to build new proteins. The heme of haemoglobin release iron. The heme is converted to bilirubin. Blood transports iron to red marrow, where it is used to produce new haemoglobin.

Blood transports bilirubin to the liver. Bilirubin is excreted as a part of the bile into the small intestine and reabsorbed from intestine into blood and excreted from the kidney in the urine.

Leukocytes (WBCs

These are spherical cells. It lacks haemoglobin. WBCs are larger than RBCs. WBCs are named according to its shape of nucleus. Nucleated polymorphic cells with normal counts ranging from 4,000-11,000 per mm cube (< 4,000 = leukopenia and >11,000 = leukocytosis). WBCs are able to slip out of capillary blood vessels (**diapedesis**) to areas of body where they are needed for an immune response. Once out of the cappilary, WBCs move through tissue spaces by **amoeboid motion**. WBCs can locate areas of tissue damage and infection by responding to chemicals released by damaged cells or other leukocytes. This process is called **positive chemotaxis**. Leukocytes are grouped into two major subdivisions: granulocytes and agranulocytes. WBCs protect body against invading micro organisms. It also removes dead cells and debris from tissue by phagocytosis.

Granulocytes

Neutrophils (Fig.70) - Neutrophils form the major part of the white blood cells. They contributed to about 40-70% of nucleated cells. They have multi lobe nucleus with granulated cytoplasm. Granules do not take up acidic or basic stains and hence named neutrophil. They are also called as polymerpho nuclear leukocytes (PMN). They are important cell found predominanty at the site of inflammation. The Process by which the circulating neutrophils enter into the tissue space is called extravasation. Neulrophils are active phagocytic cells. They have two types of granules viz. primary granules and secondary granules. Primary granules are like lysosomes, larger in size and contains peroxidase, lysozyme and hydrolytic enzymes. Secondary granules are smaller in size and contain enzymes like collagenase, lactoferrin and lysozyme. During phagocytosis, foreign materials are enclosed in a sac like structure called phagosomes. Then the primary and secondary granules are fuse together and eliminate foreign particle. Neutrophil also express a complement receptor (CD 35) and receptor for Fc portion of immunoglobulin (CD16).

Eosinophils (Fig.70) - They have a bilobed nucleus. The granules are stained with acidic dyes such as eosin and hence its name. The granules are rich in hydrolytic enzymes. They are about 2-5% of the leucocytes in the healthy individual. Increased number of eosinophils are observed in allergic patients and are called eosinophilia. Eosinophils are involved in the immunity against parasitic infections. They kill helminthes by releasing chemical mediators stored in granules as they cannot phagocytose them since they are larger organisms.

Basophils (Fig.70) : Granuels of these cells are stained with basic dyes such as methylene blue. Hence the name basophils. They contribute about 0.2 to 0.4% to the total leucocyte population. Gramules of basophil contain histamine and other vosoactive substances. They possess receptor for a unique class of immunoglobulin IgE, which mediate allergic reactions. They are responsible for immediate hyper sensitivity reactions. They have lobed nucleus.

Agranulocytes

Monocytes : They are mono nuclear phagocytic cells. They are present in blood stream. They have horse shoe shaped nucleus. Cytoplasm contains a zurophilic granules. Monocytes are converted to macrophages. They circulate in blood stream for about 8-10 hours. The mature monocyte measures 12-15 m in diameter in blood smears and represents a circulating cell derived from bone marrow promonocytes. They are the largest cell in the peripheral blood. They contain a fairly large nucleus which is ovoid, kidney- or horseshoe-shaped, often located in an eccentric position. Nucleoli are sometimes seen. The cytoplasm is pale and may contain fine azurophilic granules (lysosomes), small amounts of RER, free ribosomes, polyribososomes, and a well-developed Golgi apparatus (making lysosomes). Pinocytotic (clear) vacuoles are frequently seen in cytoplasm. Cell membrane has many finger-like microvilli. Peripheral blood has 3-8% monocytes. Monocytes belong to the monocuclear-phagocyte system. These cells leave the blood and enter the tissue, differentiating into macrophages or tissue histiocytes. **They ingest (phagocytize) and remove particulate matter, tissue debris, and infectious agents.** Histiocytes/ macrophages are present in many tissues, but are particularly numerous in the bone marrow, liver and spleen. They interact with lymphocytes and play essential role with antigen interaction of immunocompetent cells.

Macrophages

Macrophages are large, mononuclear phagocytic cells. They are derived from monocytes. Monocytes get enlarged and migrate into the tissue to become the tissue specific macrophages. Macrophage contains azurophilic granules that contains lysozyme, acid hydrolases, phosphatase, B glucouronidase, myeloperoxidase. Macrophages are actively participated in phagocytosis of various pathogens and tumour cells. Usually tissue macrophages are immobile but become active when stimulated by lymphokines. Macrophages are activated by gamma interferon. Various names of macrophages are as follows.

Histocytes	– Soft tissue macrophages
Kupfter Cells	– Liver
Osteoclasts	– Bone
Microglial cells	– Brain / Neve cell
Langerhans cells	– Epidermis
Glomeruler Mesenglial cells	– Kidney
Pulmonary alveolar macrophages	– Lungs
Macrophages	– Spleen and lympnodes

Macrophages are activated by IFN, and are effective in eliminating the engulfed pathogen and secrete inflammatory mediators. Macrophages are also called as Antigen presenting cells.

They express high level of MHC class II molecules. Activated macrophages also secrete Tumor Necrosis factor and macrophages have surface receptor for C_3 component of complement (CD 35) and Fc component of antibody (CD16)

Lymphocytes

Lymphocytes are mononuclear, non granular leucocytes. They are found in blood, lymph and lymphoid tissue such as spleen, lymph nodes, tonsils, peyer's patches etc. They are spherical or ovoid in shape. About 20% leucocytes are lymphocytes. There are two types of lymphocytes. They are small and large lymphocyte. Small lymphocyte have a large nucleus with a rim of cytoplasm. They do not contain endoplasmic reticulum. Small lymphocytes are further divided into two types based on their function that are T lymphocyte and B lymphocytes. Large lymphocytes have lower nuclear cytoplasm ratio, and are granulated. It is also called as large granulated lymphocytes (LGLs). Lymphocytes normally posses specific receptors for antigens and thus mediate specific immunity.

T Lymphocytes

It is a mono nuclear, non granular leucocytes that matures in thymus. It is also called thymocyte, thymic dependent cells. T lymphocyte have a large nucleus with a rim of cytoplasm. The surface of T lymphocyte contains certain unique group of proteins called T cell surface markets.

The following are the T cell surface markers.

Erythrocyte Receptor : It recognizes the sheep erythrocyte (SRBC)

T Cell antigen receptor (TCR). It recognizes MHC class I and II antigens

The Ia protein receptor: It recognizes immune associated protein.

Interleukin receptor: It contains IL1 and IL 2 receptor.

T lymphocytes are derived from haemopoietic stem cells of bone marrow. Thymic hormones are responsible for the formation of these cells.

There are different types of T cells. They are the sub populations or subsets of T cells. They are T helper Cells (T_H cells); T supressor cells (T_S cells); T cytotoxic cells (T_C cells) and T delayed type hypersensitivity cells (T_D cells). T cells play two important functions – effector and regulatory. Th and Ts cells are regulator cells where as the Tc cells and T_D cells are effector cells.

T helper cells : These cells help B cells and other T cells in immune response. They are regulator cells. Th cells are activated even by very small quantities of antigen which cannot activate other cells. Activated cells secrete lymphokines. These lymphokines increase the response of B cells, Tc cells and T suppressor cells. The B cells are activated to produce antibodies. Th cells secrete another one lymphokine called macrophage migration inhibition factor. Which causes the accumulation of macrophages around antigen and the activated macrophage perform phagocytosis.

T Suppressor Cells : These cells suppress the activities of B cell and other T cells. They are regulatory cells. They inhibit antibody production by B cells. They suppress the function of Th cell and Tc cells. These cells are responsible for immune tolerance by limiting the ability of the immune system to attack a persons own body tissue.

T cytotoxic cell : These cells kill micro organisms or microbes infected cells. These cells release cytotoxic substances directly into the attacked cells. These cells are lethal to tissue cells that have been invaded by viruses. T_C cells also play an important role in destroying cancer cells.

T delayed type hypersensitivity cell : They are sub population of T lymphocytes. They bring macrophages to areas where delayed hypersensitivity occurs. T_D cells secrete macrophage. Chemotoxin and macrophage migration inhibition factor. By secreting these factors the T_D cells are directly involved in delayed hypersensitivity reaction.

B Lymphocytes

The lymphocytes matures in the bursa of fabricius or bone marrow and brings about humoral immunity are called B lymphocytes. B cells are mononucleated non granulated leucocytes. They have large nucleus and a rim of cytoplasm. Surface membrane of B cell contain unique proteins called B cell surface markers. They are as follows. Ia (immune associated) protein which binds to a Ia receptor of T cell. Fc receptor to bind with the Fc fragment of the immunoglobulin. CRI and CR2–receptors of complement system. Surface immunoglobulin (IgM and IgD)-specific receptor for antigen.

The B cells one of two subsets. They are T Cell dependent cells – These cells require the help of T_H cells for the production of immunoglobulin and T cell independent cells – These cells do not require the help of T cell for immunoglobulin production. Cytokines are known to stimulate B cells to become *Plasma cell.*

Plasma Cell

Plasma cells secrets different classes of antibodies such as IgG, IgA, IgM, IgD, IgE. Some plasma cells live for 2 to 3 days while others continue to produce antibodies for several weeks. Plasma cells are large lymphocytes with a large cytoplasm with small nucleus. They have basophilic cytoplasm and eccentric nucleus with heterochromatin in a characteristic *Cart wheel arrangement*. Plasma cells are end cells. The cytoplasm is completely filled with rough endoplasmic reticulum. Plasma cells are devoid of surface receptors but rich in immunoglobulin in the cytoplasm. The immunoglobulin is localized in the spaces of the endoplasmic reticulum, where it sometimes forms distinct aggregates termed Russel's nodes.

Large Granular Lymphocytes Or Null cells or Third population cells

Apart form T cell and B cells, certain groups of lymphocytes do not express membrane bound molecules and receptors. They are larger than T cell and B cells. They are also called null cells. There are two types of null cells namely Natural killer cells and Killer cells.

Natural Killer Cells : They are a group of null cell. They have 2 or 3 large granules in the cytoplasm. Hence they are also called large granular lymphocytes. They kill the target cell without the aid of antibody or complement. So they are called antibody independent. NK cells are activated by virus infected cells through interferon. They recognize altered cell surface and brings about cytolysis and cytotoxicity.

Killer Cells (K) : They are group of null cells. They are antibody dependent cells. These cells possess Fc receptors for binding IgG. These cells can bind with cells coated with IgG antibodies and can kill them. The reaction created by K cells are called ADCC reaction (Antibody dependent cell mediated cytotoxic reaction). By this cytotoxic property, K cells can kill a variety of cells such as tumour cells, bacteria, virus, fungi and parasites.

Mast Cells

Mast cells are sessile and are found in various tissues like skin, connective tissues of various organs and mucosal epithelial tissues. These cells available in a blood stream as undifferentiated cells. Once they leave the blood and enter the tissues they become mast cells.

These cells resemble the basophils in having large number cytoplasmic granules that contain histamines and other pharmacologically active substance. Mast cells are similar to basophils in their appearance and function. There are two types of mast cell, that are mucosal mast cells and connective tissue mast cells. Serotonin and prostaglandins of mast cells brings about the inflammation.

Dendritic Cells

These cells are named because they resemble the dendrits of neuron on the cell surface extensions. They circulate in blood as immature cells and matured in tissue. Dendritic cells express high levels of both calls I and Class II MHC molecule. They are potent Antigen Presenting cells. Dendritic cells process antigen and present it to Th cells. There are different types of dendritic cells, they are,

1. Langerhans cells – found in epidermins and mucous membrane.

2. Interdigitating dendritic cells – found in T cell areas of the secondary lymphoid organ.

3. Interstitial dendritic cells – found in heart, lungs, liver, kidney etc.

4. Veiled cells – circulating dendritic cells

5. Follicullar dendritic cells – found in lymph follicles.

They help in developing memory B cells with in a follicle.

Platelets (thrombocytes)

It is also called thrombocytes. It is a component of blood. It is a Non-nucleated flat, biconvex, round or ovoid disks (2-5 m diameter) like cells. They are **derived from bone marrow megakaryocytes.** The megakaryoblast differentiates into a megakaryocyte that invaginates its cell membrane to compartmentalize the cytoplasm. Once compartmentalization is complete, the megakaryocyte ruptures into fragments (thrombocytes). The production of thrombocytes is regulated by **thrombopoietin.** Platelets produce proteins (fibrin) which aid in the clotting of blood vessels. Thrombocytopenia - number of circulating platelets are deficient. Platelets are involved in **hemostasis** (stopping bleeding). Platelets aggregate on damaged endothelium and exposed collagen, release contents of alpha and dense granules, **promote the coagulation cascade** involving plasma factors **to form a blood clot.** The formation of the platelet plug is referred to as "primary hemostasis". The clotting process is good to limit hemorrhage but can be life threatening on the wall of a coronary artery (coronary thrombosis) or when breaking free from veins of extremities and going to lungs to cause pulmonary embolism. **A deficiency of platelets is called thrombocytopenia**. Thrombocytopenia may result from a number of causes, such as immune destruction by antibodies directed against platelets, bone marrow infiltration by metastatic cancer, or primary malignancy of the bone marrow. **Thrombocytopenia may result in bruising or bleeding**.

Colour of blood

The colouring matter of blood (**hemochrome**) is largely due to the protein in the blood responsible for oxygen transport. Haemoglobin is the principal determinant of the colour of blood in vertebrates.

Functions of blood

Blood perform the following functions:

1. Blood transport oxygen from lungs to the tissues and carbondioxide from the tissues to the lungs. Thus it is responsible for the important process of respiration.

2. It transports absorbed dietary nutrients from the digestive tract to all the body tissues.

3. It transports metabolic wastes to the kidneys, lungs, skin and intestine for removal.

4. It transports various hormones and minerals.

5. It maintains normal acid - base balance in the body.

6. It regulates water balance.

7. It regulates body temperature by the distribution of body heat.

8. White blood cells and antibodies in blood provide, defense against various type of infections.

PREVENTION OF BLOOD LOSS

When a blood vessel is damaged, blood can leak in to other tissues and interfere normal metabolism or it can lost from body. Large amount of blood loss leads to death. The lose of blood is managed by three process. They are vascular spasm, platelet plug formation and blood clotting.

Vascular spasm : It is an immediate but temporary construction of a blood vessel that results when smooth muscle within the wall of vessel contracts. This constriction can close small vessels completely and stop the flow of blood through them. Chemicals also produce during vascular spasm. Platelets release thromboxanes.

Platelet plug formation : It is an accumulation of platelets that can seal up a small break in a blood vessel. It is a very important in maintaining the integrity of blood vessel. Platelets stick to the collagen exposed by damaged blood vessel. This process in called platelet adhesion. Most platelet adhesion is mediated through von Willebrand factor. It is secreted by blood vessel endothelial cells. Von Willebrand factor forms bridge between collagen and platelets. After platelet adheres to collagen, This localization of platelets to the extracellular matrix promotes collagen interaction with platelet glycoprotein VI. Binding of collagen to glycoprotein VI triggers a signaling cascade that results in activation of platelet integrins. Activated integrins mediate tight binding of platelets to the extracellular matrix. This process adheres platelets to the site of injury.

Activated platelets will release the contents of stored granules into the blood plasma. The granules include ADP, serotonin, platelet-activating factor (PAF), vWF, platelet factor 4, and thromboxane A2 (TXA2), which, in turn, activate additional platelets. The granules' activate a Gq-linked protein receptor cascade, resulting in increased calcium concentration in the platelets' cytosol. The calcium activates protein kinase C, which, in turn, activates phospholipase A2 (PLA2). PLA2 then modifies the integrin membrane glycoprotein IIb/IIIa, increasing its affinity to bind fibrinogen. The activated platelets change shape from spherical to stellate, and the fibrinogen cross-links with glycoprotein IIb/IIIa aid in aggregation of adjacent platelets (completing primary hemostasis).

Coagulation

Coagulation (also known as clotting) is the process by which blood changes from a liquid to a gel, forming a blood clot. Blood clot potentially results in hemostasis. It is the cessation of blood loss from a damaged vessel. The mechanism of coagulation involves activation, adhesion, and aggregation of platelets along with deposition and maturation of fibrin.

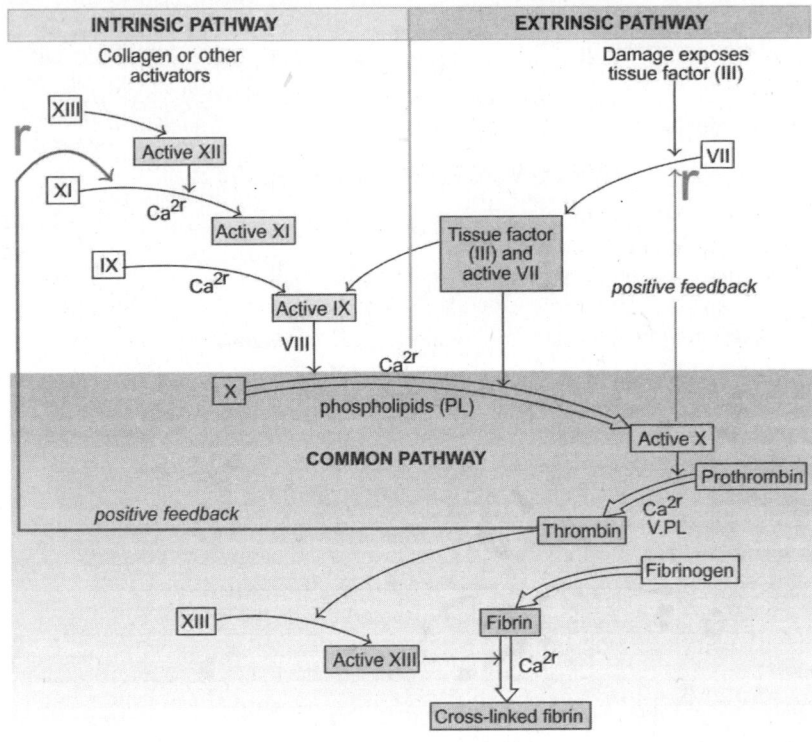

*Fig. 71 : **Blood clotting mechanism.***

Coagulation begins after an injury to the blood vessel. This damages the endothelium lining of the vessel. Leaking of blood through the endothelium initiates two processes. There are changes in platelets, and the exposure of subendothilial tissue factor to plasma Factor VII, which ultimately leads to fibrin formation. Platelets immediately form a plug at the site of injury; this is called primary hemostasis (Platelet Plug formation). Secondary hemostasis occurs simultaneously by additional coagulation factors or clotting factors beyond Factor VII respond in a complex cascade to form fibrin strands, which strengthen the platelet plug.

Platelet activation : When the endothelium is damaged, the normally isolated, underlying collagen is exposed to circulating platelets, which bind directly to collagen with collagen-specific glycoprotein Ia/IIa surface receptors. This adhesion is strengthened further by von Willebrand factor (vWF), which is released from the endothelium and from platelets; vWF forms additional links between the platelets' glycoprotein Ib/IX/V and the collagen fibrils. This localization of platelets to the extracellular matrix promotes collagen interaction with platelet glycoprotein VI. Binding of collagen to glycoprotein VI triggers a signaling cascade that results in activation of platelet integrins. Activated integrins mediate tight binding of platelets to the extracellular matrix. This process adheres platelets to the site of injury.

Coagulation cascade

The classical blood coagulation pathway : The coagulation cascade of secondary hemostasis has two initial pathways which lead to fibrin formation. The coagulation cascade is therefore classically divided into three pathways. The tissue factor and contact activation pathways both activate the "final common pathway" of factor X, thrombin and fibrin.

Tissue factor pathway (extrinsic) : The main role of the tissue factor pathway is to generate a "thrombin burst". Following damage to the blood vessel, FVII leaves the circulation and comes into contact with tissue factor (TF) expressed on tissue-factor-bearing cells (stromal fibroblasts and leukocytes), forming an activated complex (TF-FVIIa). TF-FVIIa activates FIX and FX. FVII is itself activated by thrombin, FXIa, FXII and FXa. The activation of FX (to form FXa) by TF-FVIIa is almost immediately inhibited by tissue factor pathway inhibitor (TFPI). FXa and its co-factor FVa form the prothrombinase complex, which activates prothrombin to thrombin. Thrombin then activates other components of the coagulation cascade, including FV and FVIII (which forms a complex with FIX), and activates and releases FVIII from being bound to vWF. FVIIIa is the co-factor of FIXa, and together they form the "tenase" complex, which activates FX; and so the cycle continues.

Contact activation pathway (intrinsic) : The contact activation pathway begins with formation of the primary complex on collagen by high-molecular-weight kininogen (HMWK), prekallikrein, and FXII (Hageman factor). Prekallikrein is converted to kallikrein and FXII becomes FXIIa. FXIIa converts FXI into FXIa. Factor XIa activates FIX, which with its co-factor FVIIIa form the tenase complex, which activates FX to FXa. The minor role that the contact activation pathway has in initiating clot formation. **Final common pathway** - Thrombin act on fibrinogen and converted to fibrin with the help of FXII. This inturn forms cross linked fibrin, which results in hemostasis.

The CBC (Complete Blood Count)

The CBC is a very common laboratory test. It is an evaluation of red blood cells, white blood cells, and platelets. It include total count and differential count of blood.

Introduction

The body has a remarkable capacity for maintaining homeostasis despite having to coordinate the activities of over 75 trillion cells. The principal means by which the coordination occurs is chemical messengers. Most chemical messengers are produced by a specific collection of cells are called glands. Gland is an organ consisting of epithelial cells that specialize in secretion.

Classes of chemical messengers

There are four classess of chemical messengers. They are autocrine, paracrine, neurotransmitter and endocrine chemical messengers.

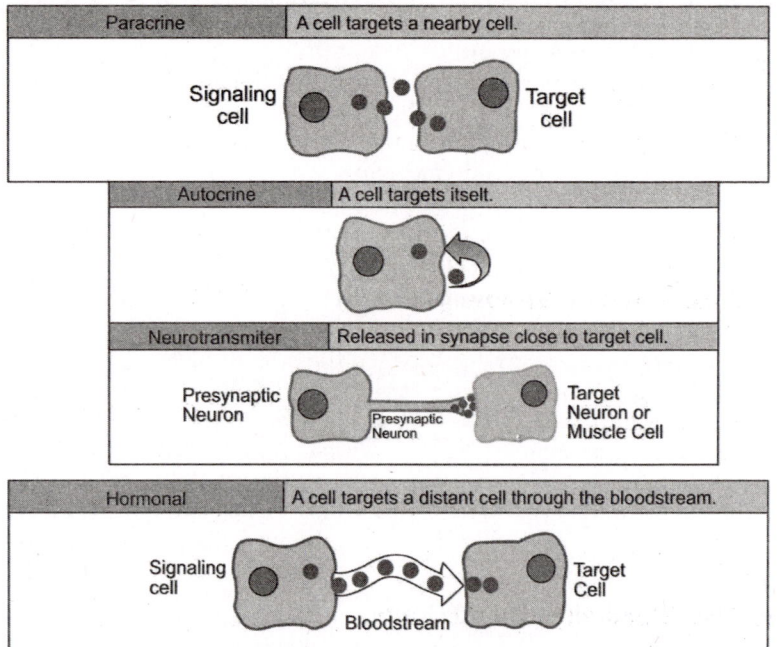

Fig. 72 : **Hormone classes.**

Autocrine is an endocrine chemical messengers are secreted by cells in a local area and influence the activity on the same cell or cell type. Eg., Prostaglandins.

Paracrine chemical messengers are produced by a wide variety of tissue and secreted into a extracellular fluid and affect the surrounding cells of different types. Eg., Histamine.

Neurotransmitters chemical messengers are produced by neurons that activate on adjacent cells. It is secreted into synaptic cleft and travels for a shorter distance. Eg. Acetyl choline.

Endocrine or Hormonal chemical messengers are secreted into blood stream by specialised cells; travels some distance to reach target tissue results in coordinated regulation of cell function.

Endocrine System

The endocrine system is made up of the endocrine glands that secrete hormones. Although there are eight major endocrine glands scattered throughout the body. Some glands also have non-endocrine regions that have functions other than hormone secretion. For example, the pancreas has a major exocrine portion that secretes digestive enzymes and an endocrine portion that secretes hormones. The ovaries and testes secrete hormones and also produce the ova and sperm. Some organs, such as the stomach, intestines, and heart, produce hormones, but their primary function is not hormone secretion. Endocrine glands have no ducts to carry secretions. Hence it is called ductless glands.

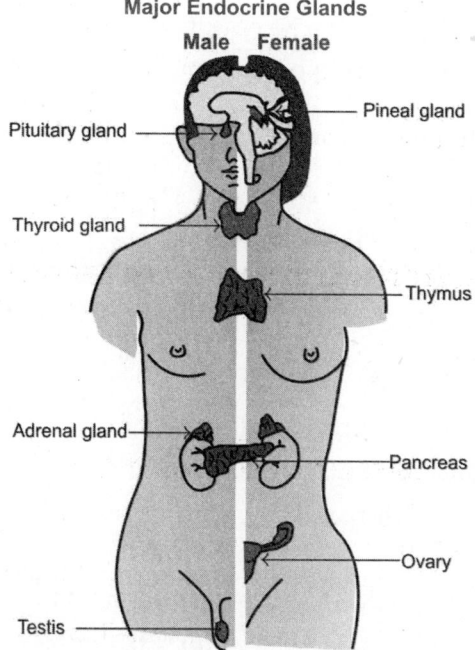

Major Endocrine Glands

Male   Female

Pituitary gland — ... — Pineal gland

Thyroid gland — ...

... — Thymus

Adrenal gland — ...

... — Pancreas

... — Ovary

Testis — ...

*Fig. 73 : **Endocrine glands.***

Main regulatory functions of endocrine system/hormone are :

1. Metabolism

2. Control of food intake and digestion.

3. Tissue development.

4. Ion regulation

5. Water balance

6. Heart rate and blood pressure regulation.

7. Control of blood glucose and other nutrients.

8. Control of reproductive functions.

9. Uterine contraction and milk release.

10. Immune system regulation.

Characters of hormone

The word hormone is derived from the Greek word hormone which means to set into motion. Hormone regulates almost every physiological functions of the body. **Hormones** are substances produced by highly specialized tissues called the "Endocrine" or "ductless glands. Hormones are chemical messengers. They are secreted by endocrine glands. They are secreted in small quantities. They are acting as a catalysts and co-enzymes. They are soluble in water. Single harmone have multiple effects. They are low molecular weight substances. Hormones are not species specific. They are under the control of neuron. They are biosynthesized. They have short half lives. They are feedback regulated by themselves, the product(s) of their action and by the central nervous system.

Hormone Receptors : Interaction between hormone and receptor forms the "hormone receptor complex". Strength of binding is expressed as the dissociation constant Kd, the concentration at which the binding sites are half-saturated.

CLASSIFICATION OF HORMONES BY RECEPTOR PROPERTIES

Group I : Hormones that bind to intracellular receptors, Eg. Glucocorticoids, mineralocorticoids, estrogens, progestins, androgens, vitamin D, thyroid, retinoic acid

Group II : Hormones that bind to cell surface receptors. Eg. Vasopresin.

CLASSIFICATION BASED ON SOLUBILITY

Based on solubility there are two types of hormones. They are lipid soluble and water soluble hormones. Lipid soluble hormones are nonpolar in nature. They are steroid hormone, thyroid hormone and fatty acid derivative hormone. These hormones are transported by binding proteins. Water soluble hormone are polar molecule. It includes protein hormone, peptide and aminoacid derived hormone. It is water soluble in nature and these hormones can dissolve in blood and circulate as free hormone. They have relatively short life span because they are easily degraded by enzymes.

CLASIFICATION BASED ON CHEMICALS

Based on chemicals hormones are classified as steroid, glycoproteins, proteins and amine hormones. Steroid hormons are derived from cholesterol. Amine hormones are secreted by gland cells of nervous system. Certain hormones are derived from amino acids, peptides and proteins.

Type	Description	Example
Amines	Hormones derived from the aminoacids tryptophan and tyrosine	Epinephrine, Norepinephrine, Thyroxin, triodothyronine.
Peptides and Proteins	Made purely from long chains of proteins	Oxytocin, Insulin, antidiuretic hormone.
Glycoprotein	Hormones made of Proteins attached with Carbohydrates	FSH and LH
Steroids	These are hormones derived from cholesterol	Testosterone, Oestrogen

Control of hormone secretions

Three types of stimuli regulate hormone release. They are humoral stimuli, neural stimuli and hormonal stimuli.

Humoral Stimuli

The term "humoral" is derived from the term "humor,". It refers to bodily fluids such as blood. A humoral stimulus refers to the control of hormone release in response to changes in extracellular fluids such as blood or the ion concentration in the blood. For example, a rise in blood glucose levels triggers the pancreatic release of insulin. Insulin causes blood glucose levels to drop, which signals the pancreas to stop producing insulin in a negative feedback loop.

Hormonal Stimuli

Hormonal stimuli refers to the release of a hormone in response to another hormone. A number of endocrine glands release hormones when stimulated by hormones released by other endocrine glands. For example, the hypothalamus produces hormones that stimulate the anterior portion of the pituitary gland. The anterior pituitary in turn releases hormones that regulate hormone production by other endocrine glands. The anterior pituitary releases the thyroid-stimulating hormone, which then stimulates the thyroid gland to produce the hormones T3 and T4. As blood concentrations of T3 and T4 rise, they inhibit both the pituitary and the hypothalamus in a negative feedback loop.

Neural Stimuli

Some endocrine secretions are solely controlled by nerve impulses. Secretion of adrenal medullary hormones secretion of neuro-hypophysial hormones and various releasing hormones of hypothalamus are under this category.

For instance, in mammals, the act of suckling of baby stimulates tactile receptors in the nipple of mother and this impulse stimulates hypothalamic cells through sensory nerve and spinal cord. Latter, hypothalamic neuro-secretion stimulates neurohypophysis for secretion of oxytocin. Oxytocin helps in secretion of milk.

PITUITARY GLAND

It is an endocrine gland. It is about the size of a pea and weighing 0.5 grams in humans. The pituitary gland is also called as hypophysis. It is named by Vesalius. It rests in a depression of the sphenoid bone inferior to the hyphothalamus of the brain. Hypothalamous is an important autonomic nervous system and endocrine control center of the brain. It is located inferior to the thalamus. It lies posterior to the optic chiasm and the gland is connected to the hypothalamus of the brain by a slender stalk called the infundibulum. There are two distinct regions in the gland. They are the anterior pituitary (adenohypophysis) and the posterior pituitary (neurohypophysis). The activity of the adenohypophysis is controlled by releasing hormones from the hypothalamus. The neurohypophysis is controlled by nerve stimulation. Pituitary gland is formed by fusion of two structures namely infundifulam and Rathkes pocket. The anterior pituitary is madeup of epithelial cells derived from the embryonic oral cavity. Cells of anterior lobe are called Chromophobes and chromophils. The posterior pituitary is an extension of brain. It is composed of nerve cells.

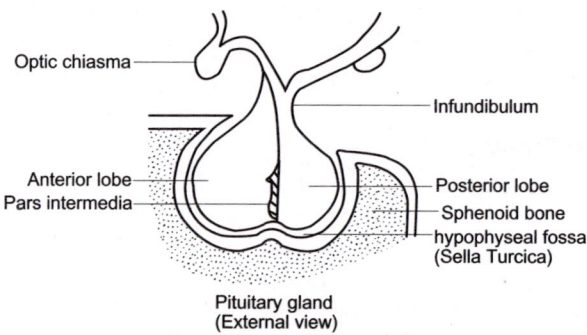

Optic chiasma
Infundibulum
Anterior lobe
Pars intermedia
Posterior lobe
Sphenoid bone
hypophyseal fossa
(Sella Turcica)
Pituitary gland
(External view)

*Fig. 74 : **Pituitary gland.***

The Anterior pituitary secretes:

- Luteinizing hormone (LH) and follicular stimulating hormone (FSH), which act on the gonads.
- Prolactin (PRL) which acts on the mammary gland
- Adrenocortiotrpic hormone (ACTH), which acts on the adrenal cortex to regulate the secretion of glucocorticoids.
- Growth hormone (GH), which acts on bone, muscle and liver.
- Thyroid stimulating hormone or thyrotropin (TSH) which stimulates the release of thyroxine (T4) and triiodothyronine (T3) from thyroid gland.

Posterior pituitary secretes:

- Antidiuretic hormone (ADH). It is also called vasopressin which controls excreted water from kidney.
- Oxytocin which controls labour.

Growth Hormone (GH) or Somatotrophic Hormone (STH)

Growth hormone stimulates growth of the entire body. It is secreted by acidophil cells of adenohypophysis (Anterior Pituitary). It is protein in nature and is formed of a straight chain polypeptide chain having about 200 aminoacids. It stimulates the growth of muscles, bones and other organs by increasing gene expression. It is also called stomatotrophic hormone (STH). It also resist protein break down during the period of food deprivation and favours lipid breakdown. Abnormality in pituitary gland showed smaller size of pituitary gland. This leads to little growth harmone secretion. This hormone drastically affects the appearance of an individual because it influences height. If there is too little growth hormone in a child, that person may become a pituitary dwarf. An excess of the hormone in a child results in an exaggerated bone growth, and the individual becomes exceptionally tall or a giant. This is called gigantism. If excess hormone is secreted after the bone length is complete, growth continues in bone diameter only. As a result, the facial features and hands become abnormally large, a condition called acromegaly. GH increases intestinal absorption of minerals. This hormone increases protein synthesis.

Gigantism

Gigantism is an abnormal growth. It is due to an excess of growth hormone during childhood. This is due to overactivity of pituitary gland. The child will grow in height, as well as in the muscles and organs. This excessive growth makes the child extremely large for his or her age. Other symptoms include: Delayed puberty, Double vision or difficulty with side (peripheral) vision, Very prominent forehead and a prominent jaw, Gaps between the teeth, Increased sweating, Irregular periods (menstruation), Large hands and feet with thick fingers and toes, Sleep problems, Thickening of the facial features, Weakness, Voice changes and increased Basal Metabolic rate.

Acromegaly

Acromegaly is a disorder that results from excess growth hormone (GH). The initial symptom is typically enlargement of the hands and feet. There may also be enlargement of the forehead, jaw, and nose. Acromegaly is typically due to the pituitary gland producing too much growth hormone. In more than 95% of cases the excess production is due to a benign tumor, known as a pituitary adenoma.

Signs and symptoms include Soft tissue swelling visibly resulting in enlargement of the hands, feet, nose, lips and ears, and a general thickening of the skin, Soft tissue swelling of internal organs, notably the heart with attendant weakening of its muscularity, and the kidneys, also the vocal cords resulting in a characteristic thick, deep voice and slowing of speech, Generalized expansion of the skull at the fontanelle, Pronounced brow protrusion, often with ocular distension (frontal bossing), Pronounced lower jaw protrusion (prognathism) with attendant macroglossia (enlargement of the tongue) and teeth spacing and Carpal tunnel syndrome

Dwarfism

It is also known as short stature. It occurs when an organism is extremely small. In humans, it is sometimes defined as an adult height of less than 4 feet 10 inches (58 in; 147 cm), regardless of sex. There are two main categories of dwarfism—disproportionate and proportionate. Disproportionate dwarfism is characterized by an average-size torso and shorter arms and legs or a shortened trunk with longer limbs. In proportionate dwarfism, the body parts are in proportion but shortened. Dwarfism can be caused by any of more than 200 conditions. Causes of proportionate dwarfism include metabolic and hormonal disorders such as growth hormone deficiency. The most common types of dwarfism, known as skeletal dysplasias, are genetic. Skeletal dysplasias are conditions of abnormal bone growth that cause disproportionate dwarfism. People with dwarfism related to growth hormone deficiency can be treated with growth hormone.

Thyroid-stimulating hormone or Thymotropin or TSH

This hormone binds to membrane bound receptors of cells of the thyroid gland and signals the thyroid to produce thyroid hormone. TSH stimulates the thyroid gland to secrete the hormone thyroxine (T4), which has only a slight effect on metabolism. T4 is converted to triiodothyronine (T3), which is the active hormone that stimulates metabolism. TSH is secreted throughout life but particularly reaches high levels during the periods of rapid growth and development. The hypothalamus, in the base of the brain, produces thyrotropin-releasing hormone (TRH). TRH stimulates the pituitary gland to produce TSH. The concentration of thyroid hormones (T3 and T4) in the blood regulates the pituitary release of TSH; when T3 and T4 concentrations are low, the production of TSH is increased, and, conversely, when T3 and T4 concentrations are high, TSH production is decreased. This is an example of a negative feedback loop.

Adrenocorticotropic hormone

This is a hormone that stimulates the adrenal cortex to get it to secrete corticosteroids. **Adrenocorticotropic hormone (ACTH)** is also known as **corticotrophin.** It is produced and secreted by the anterior pituitary gland. It is a protein hormone made up of single poly peptide chain. ACTH consists of 39 amino acids, the first 13 of which (counting from the N-terminus) may be cleaved to form α-melanocyte-stimulating hormone (α-MSH). ACTH stimulates secretion of glucocorticoid steroid hormones from adrenal cortex cells, especially in the zona fasciculata of the adrenal glands.

ACTH acts by binding to cell surface ACTH receptors, which are located primarily on adrenocortical cells of the adrenal cortex. The ACTH receptor is a seven-membrane-spanning G protein-coupled receptor. Upon ligand binding, the receptor undergoes conformation changes that stimulate the enzyme adenylyl cyclase, which leads to an increase in intracellular cAMP and subsequent activation of protein kinase A. ACTH binds to melanocytes in the skin and increase skin pigmentation. ACTH secretions increase skin colour. **Cushing's disease** due to excess adrenocorticotropic hormone (ACTH) may also result in hyperpigmentation. Deficiency of this hormone causes Addisons disease, Rheumatic Arthritis etc.,

Melanocyte-stimulating-hormone

This stimulates melanocytes. The **melanocyte-stimulating hormones**, known collectively as **MSH**. It is also known as **melanotropins** or **intermedins**. They are a family of peptide hormones and neuropeptides consisting of α-melanocyte-stimulating hormone (α-MSH), β-melanocyte-stimulating hormone (β-MSH) and γ-melanocyte-stimulating hormone (γ-MSH). This hormone is produced by cells in the intermediate lobe (Pars intermedia) of the pituitary gland. α-MSH stimulates the production and release of melanin by melanocytes in skin and hair. Production of melanin is referred to as melanogenesis. An increase in MSH will cause darker skin in humans. MSH increases in humans during pregnancy. This, along with increased estrogens causes increased pigmentation in pregnant women.

Gonadotropins

These are hormones that bind membrane bound receptors of the gonod cells. They regulate growth and developments of gonods. These are follicle stimulating hormones (FSH) and luteinizing hormones (LH) which stimulates the maturation of sex cells (Sperms or eggs) and secrete sexual hormones (testosterone or estrogen/progesterone). Gonadotropins are glyco-protein polypeptide hormones secreted by gonadotrope cells of the anterior pituitary. Gonadotropins are follicle-stimulating hormone (FSH), luteinizing hormone (LH), and placental/chorionic gonadotropins and human chorionic gonadotropin (hCG). These hormones are central to the complex endocrine system that regulates normal growth, sexual development and reproductive function. The gonadotropins act on the gonads, controlling gamete and sex hormone production. Gonadotropin is sometimes abbreviated Gn. The gonadotropins affect multiple cell types and elicit multiple responses from the target organs.

Leutinizing Hormone : It is also called Interstitial Cell Stimulating Hormone (ICSH). It is glycoprotein in nature. In cooperation with FSH, it causes the rupture of follicle and ovulation. It makes graffian follicle grow and mature. It is secreted by the anterior pituitary gland. The gonads (testes and ovaries) are the primary target organs for LH. LH stimulates the Leydig cells of the testes and the theca cells of the ovaries to produce testosterone. In female LH causes ovulation of oocytes and secretion of the hormone estrogen. In males, LH induces testes and secretes testosterone.

Folicle Stimulating Hormone (FSH) : It is a protein hormone. It is secreted by the anterior lobe of the pituitary. FSH stimulates the developments of follicles in the ovaries and sperm cells in the testes. The gonads (testes and ovaries) are the primary target organs for FSH. FSH stimulates the spermatogenic tissue of the testes and the granulosa cells of ovarian follicles, as well as stimulating production of estrogen by the ovaries.

Prolactin : It is also called Lactogenic Hormone or Luteotrophic hormone.It is secreted by anterior lobe of Pituitary gland. It is a protein hormone with several disulphide bond. Its molecular weight is 25000. Prolactin binds to membrane bound receptors in cells of breast and stimulates milk production. It helps the Corpus Lueteum in the secretion of progesterone, in cooperation with LH.

Antidiuretic hormones : It is secreted by posterior lobe of Pituitary gland. It is also called vasopressin. Its molecular weight is 1100. Antidiuretic hormones binds to membrane bound receptors and increase water reabsorption. Antidiuretic hormone promotes the reabsorption of water by the kidney tubules, with the result that less water is lost as urine. This mechanism conserves water for the body. Insufficient amounts of antidiuretic hormone cause excessive water loss in the urine. ADH can cause blood vessel to constrict when releasing in large amounts. A lack of ADH causes diabetic insipidus. It is a production of large amount of dilute urine. This leads to increase in the concentration of body fluid and loss of important electrolytes.

Oxytocin : It is also called Pitocin. It is secreted by neuro hypophysis of Pituitary gland. In Greak oxytocin means, Quick Child Birth. It is a protein formed of 9 Aminoacids. Oxytocin binds to membrane bound receptors and increase contraction of smooth muscle in the uterus. These uterine contractions happen during childbirth and after birth this hormone tells the breast tissue to eject milk into those lactiferous sinuses.

THYROID

The **thyroid gland** or the **thyroid** is an endocrine gland. It was described by Thomas Wharton. The thyroid is small, butterfly-shaped gland located in the front of the neck. It consists of two lobes connected by an isthmus. Lobes of thyroid are formed of numerous follicles. Each follicle is covered by basement membrane. In the center each follicle is filled with protein thyroglobin. The cavity is lined by a layer of cuboidal epithelial cells. Thyroid follicles consists of two types of cells. They are principal cells and Parafollicular or C cells. It is one of the largest endocrine gland. It appears more red than surrounding environment. It is surrounded by a connective tissue capsule. The thyroid weighs 25 grams in adults, with each lobe being about 5 cm long, 3 cm wide and 2 cm thick, and the isthmus about 1.25 cm in height and width. The gland is usually larger in women, and increases in size in pregnancy. The thyroid gland secretes thyroid hormones. The thyroid hormones are triiodothyronine (T_3) and thyroxine (T_4) and Calcitonin. T3 and T4 are created from iodine and tyrosine. Triiodothyronine (T_3) is synthesized in lesser quantities and conver ted to thyroxine. Calcitonin plays a role in calcium homeostasis. Hormonal output from the thyroid is regulated by thyroid-stimulating hormone (TSH) secreted from the anterior pituitary gland, which itself is regulated by thyrotropin-releasing hormone (TRH) produced by the hypothalamus.

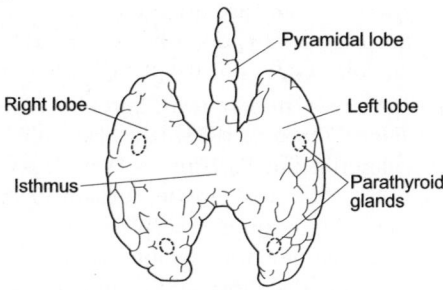

*Fig. 75 : **Thyroid gland.***

Thyroid hormone

It is a protein hormone. Thyroid hormone is synthesized and stored within numerous thyroid follicles. Each follicle is filled with protein thyroglobin, to which thyroid hormones are attached. Throid is regulated by TSH from hypothalamus and pituitary gland. It involved in growth improvement. It increases Basal Metabolic rate. It increases absorption of monosaccharides. Excessive production of the thyroid hormones is called hyperthyroidism. It is most commonly a result of Graves' disease, a toxic multinodular goitre, a solitary thyroid adenoma or inflammation. An underactive thyroid gland results in hypothyroidism. Typical symptoms are abnormal weight gain, tiredness, constipation, heavy menstrual bleeding, hair loss, cold intolerance, and a slow heart rate. The most common cause of hypothyroidism are iodine deficiency and autoimmune disease Hashimoto's thyroiditis.

Goiter : An enlarged thyroid gland is called a goitre. Goitres are present in some form in about 5% of people. It due to iodine deficiency, autoimmune diseases like Grave's disease and Hashimoto's thyroiditis, inflammation, and infiltrative disease such as sarcoidosis and amyloidoss. Sometimes no cause can be found, a state called *"simple goitre"*.

Inflammation : Inflammation of the thyroid is called thyroiditis. Inflammed thyroids may cause symptoms of hyperthyroidism or hypothyroidism. Two types thyroiditis are Hashimoto's thyroiditis and postpartum thyroiditis. Hashimoto's thyroiditis is an autoimmune disorder in which the thyroid gland is infiltrated by the lymphocytes B-cell and T-cells. Postpartum thyroiditis occurs in some females following childbirth. After delivery, the gland becomes inflamed and the condition initially presents with a period of hyperthyroidism followed by hypothyroidism and, usually, a return to normal function.

Cretinism : Infants with thyroid hormone deficiency can manifest problems of Cretinism. It is a problem of physical growth and development as well as brain development. Child indicates stunded growth. Bones and teeth are deformed. Fingers are club like. The skin is rough, thick and dry. The belly is pot like. They often showed constipation. Face appears idiotic look. Cretins are idiot. Tongue is protruding. Appetite is reduced. Body temperature reduced. Hairs are scanty on the skin. They are often def and dumb. Sex glands are retarded. Children with congenital hypothyroidism are treated supplementally with levothyroxine, which facilitates normal growth and development.

Graves' disease : Graves' disease is an autoimmune disorder. It was described by Robert James Grace in 1835. It is also called Basdows disease because symptoms are described by Basedow in 1940. It is the most common cause of hyperthyroidism. In Graves' disease, auto-antibodies are developed against the thyroid stimulating hormone receptor. These antibodies activate the receptor, leading to development of symptoms, which include a goiter, cold intolerance, weight loss, diarrhoea and palpitations. Other symptoms are protruded eye ball, osteoporosis, raised blood sugar and iodine, increased Basal Metabolic rate.

PARATHYROID GLANDS

Parathyroid glands are small glands of the endocrine system. It is located in the neck behind the thyroid. There are 4 parathyroid glands each the size of a grain of rice. It is surrounded by a connective tissue capsule. The glands control the calcium in the human bones and in the blood. Parathyroids are NOT related to the thyroid. Parathyroid glands secretes Parathyroid hormone **(PTH)**.

Parathyroid hormone (PTH) : It is a protein hormone made up of only one poly peptide. It increases serum calcium level. It increases renal tubular reabsorption of calcium. It reduces the secretion of calcium in mammary gland. It also controls intracellular deposition of phosphate. Vitamin D is not metabolized in the absence of PTH. PTH regulates calcium homeostasis. Hyperparathyroidism is the state in which there is excess parathyroid hormone circulating. This may cause bone pain and tenderness, due to increased bone resorption. Due to increased circulating calcium, there may be other symptoms associated with hypercalcemia, most commonly dehydration.

Renal disease may lead to hyperparathyroidism. The state of decreased parathyroid activity is known as hypoparathyroidism. Hypoparathyroidism will occur after surgical removal of the parathyroid glands.

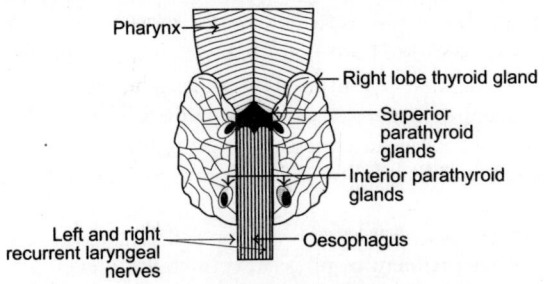

Fig. 76 : Parathyroid gland.

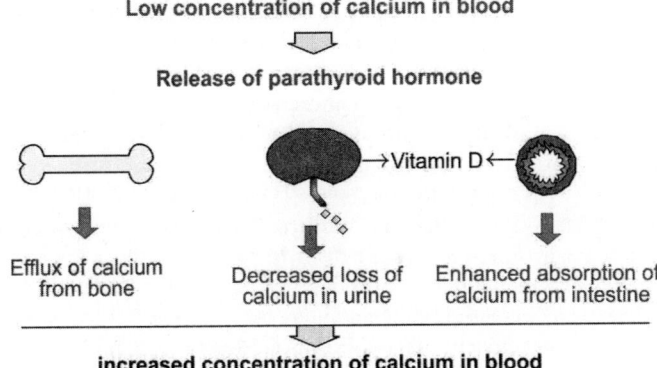

Low concentration of calcium in blood

Release of parathyroid hormone

→Vitamin D←

| Efflux of calcium from bone | Decreased loss of calcium in urine | Enhanced absorption of calcium from intestine |

increased concentration of calcium in blood

Fig. 77 : Mechanism of Para thyroid hormone.

PANCREAS

The **pancreas** is a glandular organ in the digestive system. It is located in the abdominal cavity behind the stomach. It is an endocrine gland producing several important hormones, including insulin, glucagon, somatostatin and pancreatic polypeptide.

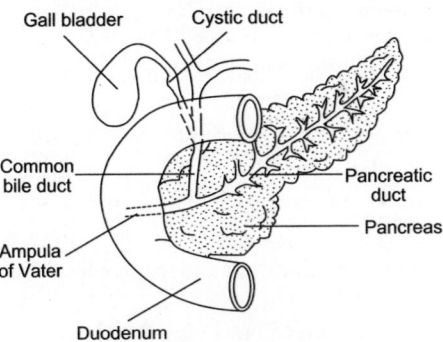

Gall bladder   Cystic duct

Common bile duct

Pancreatic duct

Pancreas

Ampula of Vater

Duodenum

Fig. 78 : Pancreas.

The pancreas is also a digestive organ, secreting pancreatic juice containing digestive enzymes. Insulin and Glucagon controls glucose level in the blood. The pancreas is involved in blood sugar control and metabolism within the body. Islets of Langerhans are present in the pancreas. They have four types of cells. They are cells secrete glucagon (increase glucose in blood), cells secrete insulin (decrease glucose in blood), cells secrete somatostatin (regulates/stops cells) and PP cells, or γ-cells, secrete pancreatic polypeptide.

INSULIN

It is a hormone secreted by cells of islets of langerhans. It is a poly peptide. It lowers blood sugar. It makes cells more permeable to glucose. It enhances glucose oxidation in cell.

It induces glycogen synthesis. It increases potassium uptake, decreases gluconeogenesis and glycogenolysis, increases lipid synthesis, increases esterification of fatty acids, decreased lipolysis, decreases proteolysis. It also increases amino acid uptake and decreases renal sodium excretion.

Diabetes Mellitus : Diabetes mellitus (DM), commonly referred to as diabetes. It is a group of metabolic disorders in which there are high blood sugars levels over a prolonged period. Symptoms of high blood sugar include frequent urination, increased thirst, and increased hunger. Acute complications can include diabetic ketoacidosis, hyperosmolar hyperglycemic state, or death. Serious long-term complications include cardiovascular disease, stroke, chronic kidney disease, foot ulcers, and damage to the eyes. Diabetes is due to either the pancreas not producing enough insulin or the cells of the body not responding properly to the insulin produced.

There are three main types of diabetes mellitus :

Type 1 DM results from the pancreas's failure to produce enough insulin. This form was previously referred to as "insulin-dependent diabetes mellitus" (IDDM) or "juvenile diabetes".

Type 2 DM begins with insulin resistance, a condition in which cells fail to respond to insulin properly. This form was previously referred to as "non insulin-dependent diabetes mellitus" (NIDDM) or "adult-onset diabetes". The most common cause is excessive body weight and not enough exercise.

Gestational diabetes is the third main form and occurs when pregnant women without a previous history of diabetes develop high blood sugar levels.

Prevention and treatment involve maintaining a healthy diet, regular physical exercise, a normal body weight, and avoiding use of tobacco.

Type 1 DM must be managed with insulin injections.

Type 2 DM may be treated with medications with or without insulin. Insulin and some oral medications can cause low blood sugar.

ADRENAL GLAND

Man contains two adrenal gland. The **adrenal glands** are also known as **suprarenal glands.** They are endocrine glands that produce a variety of hormones. It looks like a cock hat. They are found above the kidneys. Each gland has an outer cortex and inner medulla. The adrenal cortex itself is divided into three zones: zona glomerulosa, the zona fasciculata and the zona reticularis.

Adrenal cortex

Hormones secreted by Adrenal cortex are called corticosteroids. Cortex secretes three classess of hormone. They are mineralocarticoids, glucocorticoids and androgens.Mineralocarticoids are a steroid hormone secreted by adrenal cortex. It helps regulate blood volume and blood K⁺ and Na+ levels.Aldosterone is a major class of hormone of this group. It binds to receptors in the kidney. Blood levels of K⁺ and Na⁺ directly affects the adrenal cortex to influence aldosterone secretion. Its secretion is influenced when blood K⁺ level increases and Na⁺ level decrease. Changes in blood pressure indirectly affect aldosterone secretion.

Second hormone secreted by middle layer of adrenal hormone is *glucocorticoids. It hels to regulate blood nutrient level. Main hormone is cortisol.* Cortisol which affects glucose metabolism and immune status. Cortisol increase the breakdown of protein and lipid and increase their conversion to energy. Cortisol reduce inflammation and immune response. Another hormone secreted by inner layer of adrenal cortex is called androsterone. It is composed of androgens. They stimulate male sexual character. Small amount of androgen is indicated in male and female children. Androstenedione which is an anabolic male hormone and/or substrate for estrogen. Aldosterone which increase blood volume by reabsorption of sodium in kidneys.

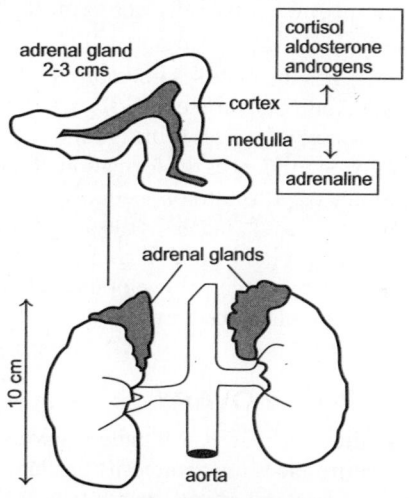

*Fig. 79 : **Adrenal gland.***

ADRENAL MEDULLA

It is a central part of adrenal gland.It is derived from neural crest cells.It secretes two types of hormone. They are Epinephrine and Norepinephrine.It produces these hormones in response to stimulation by sympathetic nervous system.Stress and low blood glucose can cause increased sympathetic stimulation.These two hormones are refered as fight or flight hormone because there role in preparing body for vigourous physical activity.Epinephrine and Norepinephrine act as neurotransmitters. Major effect of Epinephrine and Norepinephrine are increases in the breakdown of glycogen to glucose in liver, increases the release of glucose in blood and also tissues, increase the release of fatty acids from adipose tissue, increased heart rate, which causes blood pressure to rise, increased metabolic rate, blood sugar level increased, It raises oxygen consumption and respiratory rate increased rapidly.

GONAD

A gonad or sex gland or reproductive gland is an endocrine gland that produces the gametes. In the female, reproductive cells are the egg cells. In the male, the reproductive cells are the sperm. The male gonad, the testicle, produces sperm in the form of spermatozoa. The female gonad, the ovary, produces egg cells. Both of these gametes, are haploid germ cells.Gonods are under the control of the pituitary gland.

TESTES

It is a cytogenic organ. It is a **reproductive gland.** It is an endocrine gland. It produces a reproductive cells are called the sperm. Sperm is produced from spermatozoa. Sperm is a haploid germ cells.It is under the control of the pituitary gland. In the testes, interstitial cells secrete mostly testosterone, which is a type of androgen.

Androgens : Androgen is the male sex hormones. It includes testosterone and dihydrotestosterone. Androgens are secreted by testis, adrenal cortex, ovary and placenta. It is a derivative of sterol. It brings about development of male genetalia. The **testosterone** influences the formation of sperm in seminiferous tubules. Testosterones are responsible for secondary sex characteristics like pubic hair, axillary hair, facial hair. It increases the development of reproductive organs like penis, prostrate gland etc., during puberty and maintaining them into adulthood. It causes muscular development. It increases deposition of Calcium in bones. It stimulates erythropoiesis. It increases thickness of skin.

OVARY

Ovary is a cytogenic organ. It is a female **reproductive gland.** It is an endocrine gland that produces the gametes.It produces ovum or egg. Egg is a haploid germ cells. Ovary are under the control of the pituitary gland. *Ovary secretes hormone called* Estrogens. In the ovaries, interstitial cells are called **follicle cells** (some turn into **corpus luteum**) and secrete a different version of androgen called estrogen, relaxin and progesterone. **Progesterone** (P for Prepare) prepares the uterus for pregnancy.

Estrogen : Estrogen is the female sex hormones. It include Estradiol and Estrone. **Estrogen** helps produce pubic and axillary hair (no facial hair). It develop and maintain the sexual organs. It is a steroid hormone. It is secreted by ovary. Estrogens inhibit the secretion of FSH. It stimulates the secretion of ACTH. It increases total body protein. It is responsible for the development of secondary sexual characters in female. It causes increased deposition of fat in subcutaneous tissue.

Progesterone

It is a female sexual hormone. It is steroid in nature. It is secreted by corpus luteum. It prepares the uterus for pregnancy. It is responsible for premenstrual changes in the body. It helps in implantation of fertilized ovum in uterus. It completes the development of breast.

PINEAL GLAND

The pineal gland is also called pineal body or epiphysis cerebri. It is formed of paranchymal cells and interstitial cells. It is a small cone-shaped structure that extends posteriorly from the third ventricle of the brain. The pineal gland consists of portions of neurons, neuroglial cells, and specialized secretory cells called pinealocytes. The pinealocytes synthesize the hormone melatonin. Melatonin affects reproductive development and daily physiologic cycles. It affects the activities of ovary.

PLACENTA

The placenta is an organ that connects the developing fetus to the uterine wall to allow nutrient uptake, provide thermo-regulation to the fetus, waste elimination, and gas exchange via the mother's blood supply. It also helps in fight against internal infection and produce hormones to support pregnancy. The placenta provides oxygen and nutrients to growing babies and removes waste products from the baby's blood. The placenta attaches to the wall of the uterus, and the baby's umbilical cord develops from the placenta. The umbilical cord is what connects the mother and the baby. Placentas are a defining characteristic of placental mammals, but are also found in some non-mammals with varying levels of development

Important hormones of placenta are as follows : The first hormone released by the placenta is called the human chorionic gonadotropin hormone. This is responsible for stopping the process at the end of menses when the Corpus luteum ceases activity and atrophies. Progesterone helps the embryo implant by assisting passage through the fallopian tubes. It also affects the fallopian tubes and the uterus by stimulating an increase in secretions necessary for fetal nutrition. Estrogen is a crucial hormone in the process of proliferation. This involves the enlargement of the breasts and uterus, allowing for growth of the fetus and production of milk. Estrogen is also responsible for increased blood supply towards the end of pregnancy through vasodilation. Human placental lactogen is a hormone used in pregnancy to develop fetal metabolism and general growth and development.

CHAPTER 10

PLANT HORMONE

Introduction

Growth hormone of plant are organic molecule produced by the plant for regulating growth and other physiological functions. First idea of plant hormones was given by Von Sachs and named as "organ forming substance." The term hormone was given by Starling and Phytohormone by Thieman.

AUXIN

Auxin is a phytohromone. It is a organic compound. It is a first plant hormone discovered by F.W. Went. The term auxin was introduced by Kogland Haegen Smith in 1931.

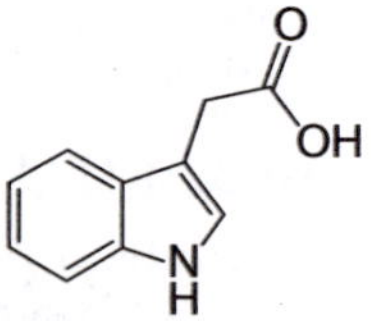

*Fig. 80 : **Auxin.***

Auxins are widely distributed throughout plant body. There are three types of auxins. They are auxin A,B and hetero auxin. Heteroauxin is also called Indole-3-acetic acid (IAA). It is an universal in occurrence. Auxin a is called auxentriolic acid. It occurs in buds and growing leaves. Auxin b is called auxenolonic acid. IAA is the main auxin in most plants. Compounds which serve as IAA precursors may also have auxin activity (e.g., indole acetaldehyde). Some plants contain other compounds that display weak auxin activity (e.g., phenylacetic acid). IAA may also be present as various conjugates such as indole acetyl aspartate. Several synthetic auxins are also used in commercial applications aspartate. IAA is synthesized from tryptophan or indole primarily in leaf primordia and young leaves, and in developing seeds. IAA transport is cell to cell, mainly in the vascular cambium and the procambial strands, but probably also in epidermal cells. Transport to the root probably also involves the phloem.

Functions

- Auxin stimulates cell enlargement and stem growth.
- Auxin stimulates cell division in the cambium.
- Auxin stimulates differentiation of phloem and xylem.
- Auxin stimulates root initiation on stem cuttings, and also the development of branch roots and the differentiation of roots in tissue culture.
- Auxin mediates the tropistic (bending) response of shoots and roots to gravity and light.
- The auxin supply from the apical bud represses the growth of lateral buds. This is called apical dominance.
- Auxin delays leaf senescence.

- Auxin may inhibit or promote leaf and fruit abscission.
- Auxin induces fruit setting and growth in some fruit.
- Auxin delays ripening.
- Auxin promotes flowering.
- Growth of flower parts are stimulated by auxin.
- Auxin promotes femaleness in dioecious flowers.

Auxin	Uses
Napthalene Acetic acid (NAA)	Induces rooting of cuttings; Prevents Sprouting in potatoes; Induces flowering; helps natural fruit settings
Indole Propionic acid (IPA)	Induces rooting of cuttings; helps natural fruit settings
Indole Butyric Acid (IBA)	Induces rooting of cuttings
2,4-dichlorophenoxy Acetic acid (2,4,D)	Selective weed killer; Causes Flowering; Prevents fruit drops.
2,3, 6-Trichlorobenzoic acid	Powerfull weed killer
2,4, 6 Trichlorobenzoic acid	Powerfull weed killer

GIBBERELLINS (GAs)

Yabuta and Sumiki in 1938 extracted a crystalline substance from the Gibberella fungus and named as Gibberellin. Gibberellin is acidic in nature. It posses a Cibben ring structure that are able to overcome genetic dwarfism in plants. 100 type of Gibberellins are known. They are named as GA_1, GA_2, GA_3, ... GA_{100}. GA_3 [$C_{19}H_{26}O_6$] is representative of all Gibberellins. First discovered Gibberellins from higher plants was GA_1.

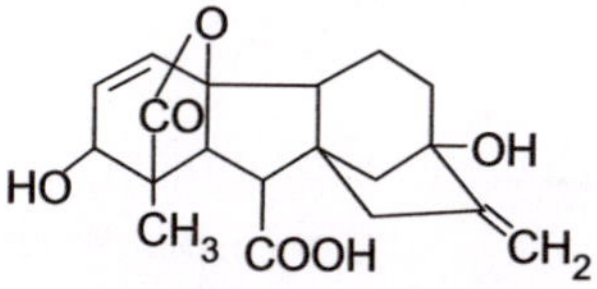

Fig. 81 : *Gibberellic acid (GA3).*

GA found in all groups of plants. It acts as a flowering hormone in angiosperms. Biosynthesis of gibbereilin takes places by Mevalonic acid pathway.

Giberellin A has following functional groups. They are one carboxyl group, one ethylenic double bond, two alcoholic hydroxyl bonds, one saturated lactone and one methyl group (**Fig.81**). The gibberellins (GAs) are a family of compounds based on the *ent*gibberellane structure. GAs are synthesized from glyceraldehyde-3-phosphate, via isopentenyl diphosphate, in young tissues of the shoot and developing seed. Some GAs are probably transported in the phloem and xylem.

Functions

- GA1 causes hyperelongation of stems by stimulating both cell division and cell elongation .

- GAs cause stem elongation in response to long days.

- GAs can cause seed germination in some seeds that normally require cold (stratification) or light to induce germination.

- GA stimulates the production of numerous enzymes, notably - amylase, in germinating cereal grains.

- GA induce fruit setting and growth.

- GA induce maleness in dioecious flowers.

CYTOKININS (CKs)

Cytokinins are plant hormone promoting cytokinesis in plants. The term cytokinin is proposed by Lethem in 1963. CKs are adenine derivatives characterized by an ability to induce cell division in tissue culture. The most common cytokinin base in plant is zeatin. Cytokinins also occur as ribosides and ribotides. CK biosynthesis is through the biochemical modification of adenine. It occurs in root tips and developing seeds. CK transport is via the xylem from roots to shoots.

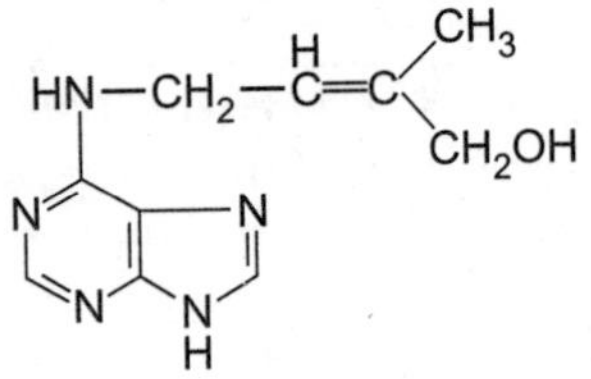

Fig.82 : Cytokinins

Functions

- It stimulates cell division and promotes cell elongation.

- They are effective in seed dormancy in lettuce, tobacco etc.,

- Exogenous applications of CKs induce cell division in tissue culture

- CKs promote shoot initiation.

- In moss, CKs induce bud formation.

- CKs can be employed successfully to induce flowering in short day plants.

- CKs delay leaf senescence.

- CKs may enhance stomatal opening in some species.

- The application of CK leads to an accumulation of chlorophyll and promotes the conversion of etioplasts into chloroplasts.

ABSCISIC ACID (ABA)

It is a plant hormone. It was discovered by Eagles and Wareing in 1963. This hormone induces dormance. So it is also called dormin. Ohkuma et al., in 1965 named this hormone as abscisin II. Chemically this hormone is sesquiterpene. One of the main functions is the regulation of stomatal closure. ABA is synthesized from glyceraldehyde-3-phosphate via isopentenyl diphosphate and carotenoids in roots and mature leaves, particularly in response to water stress. Seeds are also rich in ABA which may be imported from the leaves or synthesized *in-situ*. ABA is exported from roots in the xylem and from leaves in the phloem.

Functions

- Water shortage brings about an increase in ABA which leads to stomatal closure and reduce water loss.
- ABA inhibits shoot growth. This may represent a response to water stress.
- ABA induces storage protein synthesis in seeds.
- ABA counteracts the effect of gibberellin on -amylase synthesis in germinating cereal grains.
- ABA affects the induction and maintenance of some aspects of dormancy in seeds.
- Increase in ABA in response to wounding induces gene transcription, notably for proteinase inhibitors, so it may be involved in defense against insect attack.

Introduction

Pigments are biological substance found in both plant and animals. A pigment is defined as any substance capable of absorbing light. Photosynthetic pigments are found in plants. Colour is the property of electromagnetic radiation with a wavelength between 300 and 800 nm.

Importance of Pigments

Plant pigments are important in signalling, in attracting pollination and dispersal agents, and repelling herbivores. Plant pigments are important clues to humans and other herbivorous animals in helping identify plant parts such as fruit, leaves, stems, roots or tubers and determine stages of plant development such as ripeness or overall senescence. It was realized early in this century that many of these pigments play a positive role in human health. Plant pigments are obviously physiologically important and recent research suggests novel protective mechanisms, both photo-protective and anti-oxidative. Plant pigments are economically important, in determining the colours and patterns of attractive flowers and valuable fruits. They are also important nutritionally, with an understanding of emerging new roles in nutrition and in aiding health.

CHLOROPHYLL

Chlorophyll is a green photosynthetic pigment. It is found in plants, algae and cyanobacteria. Chlorophyll is vital for photosynthesis. It allows plants to absorb energy from light. Chlorophylls are greenish pigments which contain a porphyrin ring. Chlorophyll was first isolated and named by Joseph Bienaimé Caventou and Pierre Joseph Pelletier in 1817. Chlorophyll absorbs mostly in the blue and to a lesser extent red portions of the electromagnetic spectrum, hence it is intense green in colour. Chlorophyll molecules are specifically arranged in and around photosystems that are embedded in the thylakoid membranes of chloroplasts. In plants, chlorophyll may be synthesized from succinyl-CoA and glycine. The immediate precursor to chlorophyll a and b is protochlorophyllide. Chlorophyll is a stable ring-shaped molecule around which electrons are free to migrate. Because the electrons move freely, the ring has the potential to gain or lose electrons easily and thus the potential to provide energized electrons to other molecules. This is the fundamental process by which chlorophyll "captures" the energy of sunlight. In chlorophyll, the central ion is magnesium and the large organic molecule is a porphyrin. The porphyrin contains four nitrogen atoms that form bonds to magnesium in a square planar arrangement. Chlorophyll is one of the most important chelates in nature. It is capable of channelling the energy of sunlight into chemical energy through the process of photosynthesis. In photosynthesis, the energy absorbed by chlorophyll transforms carbon dioxide and water into carbohydrates and oxygen.

$$CO_2 + H_2O \rightarrow (CH_2O) + O_2$$

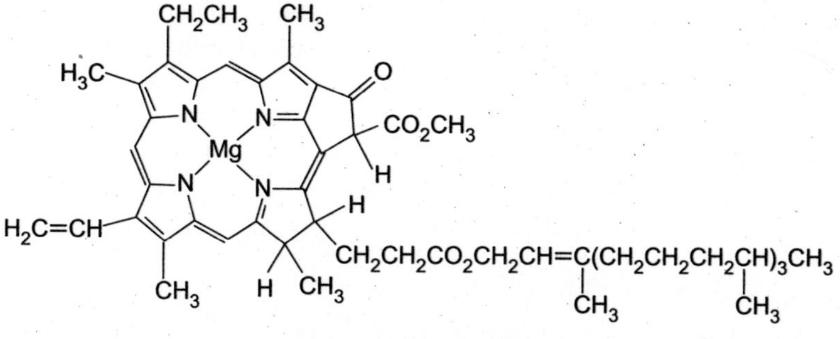

Fig. 83 : *Chlorophyll.*

In the photosynthetic reaction, carbon dioxide is reduced by water. Chlorophyll assists this reaction. When chlorophyll absorbs light energy, an electron in chlorophyll is excited from a lower energy state to a higher energy state. In this higher energy state, this electron is more readily transferred to another molecule. This starts a chain of electron-transfer steps, which ends with an electron transferred to carbon dioxide.

Types of Chlorophyll : There are several kinds of chlorophyll. The most important being chlorophyll "a". This is the molecule which makes photosynthesis possible, by passing its energized electrons on to molecules which will manufacture sugars. All plants, algae, and cyanobacteria which photosynthesize contain chlorophyll "a". A second kind of chlorophyll is chlorophyll "b", which occurs only in "green algae" and in the plants. A third form of chlorophyll which is common is called chlorophyll "c", and is found dinoflagellates. Chlorophyll d is found in many rhodophyceae members. Chlorophyll e is found only in two genera of xanthophyceae. Bacterohlorophyll is found in photosynthetic bacteria.

Chlorophyll is dissolved in organic solvents like acetone. Chlorophyll a is blue black in colour. Chlorophyll b is dark green in colour.

CAROTENOIDS

Carotenoids form one of the most important classes of plant pigments. They play a crucial role in defining the quality parameters of fruit and vegetables. They are found principally in plants, algae, and photosynthetic bacteria. They also occur in some non-photosynthetic bacteria, yeasts, and molds. Animals are unable to synthesise carotenoids de novo, and rely upon the diet as a source of these compounds.

Carotenoids are a group of naturally occurring phytochemicals that are structurally related to alpha-carotene. They are antioxidants with a number of health benefits and mostly found in fruits and vegetables. The combined concentrations of lutein and zeaxanthin are generally higher than the concentrations of the carotenes in most food sources. Alpha-Carotene and beta-carotene are precursors of vitamin A. Beta-carotene is the most common form of carotene and belongs to the group of terpenoids. Pure beta-carotene is red to purple coloured oil. It is not soluble in water. Beta-carotene occurs in coloured fruits and vegetables such as mango, apricot, sweet potatoes, carrots, kale, broccoli, spinach, turnip greens, winter squash and collard greens.

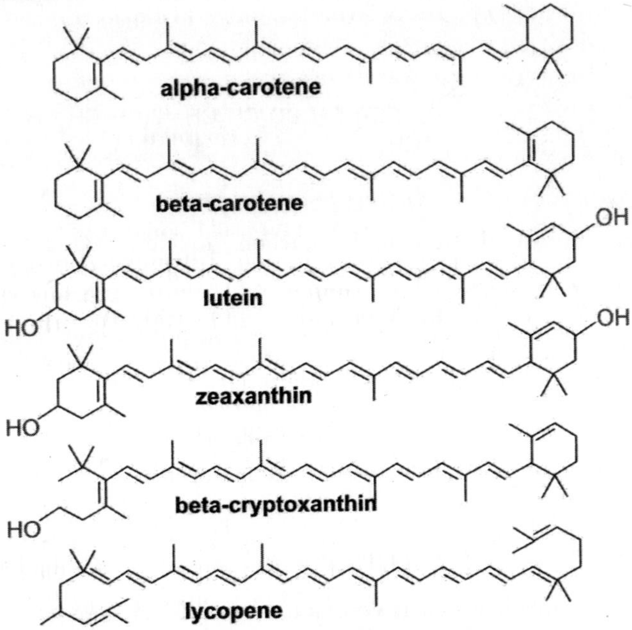

Fig. 84 : *Carotenoid.*

Classification

Carotenoids are classified according to the structure as follows:

1. The hydrocarbon carotenoids which are known as carotenes. Eg., β-carotene

2. The oxygenated carotenoids which are derivatives of these hydrocarbons known as xanthophylls. Eg., zeaxanthin and lutein.

Structure

Carotenoids are lipid-soluble C40 tetraterpenoids. The majority carotenoids are derived from a 40-carbon polyene chain, which could be considered the backbone of the molecule. This chain may be terminated by cyclic end-groups (rings) and may be complemented with oxygen-containing functional groups.

Biosynthesis

Carotenoid biosynthesis takes place in plastid.

1. The carotenoid pathway is catalyzed by phytoene synthase (PSY), resulting in the condensation of two C-20 geranylgeranyl diphosphate (GGPP) molecules to form phytoene.

2. Four desaturation reactions, two each catalyzed by the membrane associated phytoene desaturase (PDS) and ξ-carotene desaturase (ZDS), result in the formation of the pink lycopene from the colourless phytoene.

3. The cyclization of lycopene represents a branch point in the pathway, and two products can be formed depending on the position of the double bond on the cyclohexane ring.

4. The α and β-carotenes are the precursors for the xanthophylls, which are oxygenated carotenoids generated by β- and ε-ring specific hydroxylases.

5. β-carotene is converted to zeaxanthin by the carotenoid β-ring hydroxylases (HYDB), encoding a nonheme diiron enzyme for which there are two genes in Arabidopsis. The hydroxylation of the ε-ring is carried out by the carotenoid ε-ring hydroxylase (HYD-E), a cytochrome

Function

Carotenoids are essential for plant life. It provides important photoprotective functions during photosynthesis; light harvesting and prevention of photo-oxidative damage, and serving as precursors for the biosynthesis of the phytohormone abscisic acid (ABA).

They have a role in attraction of pollinators.

Beta-carotene has received a lot of attention as potential anti-cancer and anti-aging phytochemical. Beta-carotene is a powerful antioxidant, protecting the cells of the body from damage caused by free radicals. Beta-carotene is one of the many carotenoids that our body can convert into vitamin A (retinol).

Beta-carotene acts as an anti-cancer agent through its antioxidant property.

It reduces UV-induced redness of the skin and improves melasma. Beta-carotene is often use in supplements or topical creams to protect our skin.

Beta-carotene may also help to prevent arteriosclerosis by inhibiting the oxidation of lipids.

PHYCOBILINS

Phycobilins are water-soluble pigments. It is found in the stroma of chloroplast organelles. They are present only in Cyanobacteria and Rhodophyta. The two classes of phycobilins include phycocyanin and phycoerythrin. Phycocyanin is a bluish pigment found in primarily cyanobacteria (blue-green algae) to aid in absorption of light in photosynthesis. Phycoerythrin is a pigment found in Rhodopyta (red algae) that is responsible for its characteristic red colour. It is an accessory pigment that allows red algae to carry out photosynthesis in deep water.

There are four types of phycobilins. They are Phycoerythrobilin (red), Phycourobilin (orange); Phycoviolobilin found in phycoerythrocyanin and Phycocyanobilin (also known as phycobiliverdin – blue). In chemical terms, phycobilins consist of an open chain of four pyrrole rings (*tetrapyrrole*) and are structurally similar to the bile pigment bilirubin, which explains the name. Phycobilins are also closely related to the chromophores of the light-detecting plant pigment phytochrome, which also consist of an open chain of four pyrroles. Then the potential applications and uses of these pigments, pigment-protein complexes and related products by the food industr, by the health industry. It has antioxidant, anticancer, anti-inflammatory, immunomodulatory, hepatoprotective, nephroprotective and neuroprotective effects.

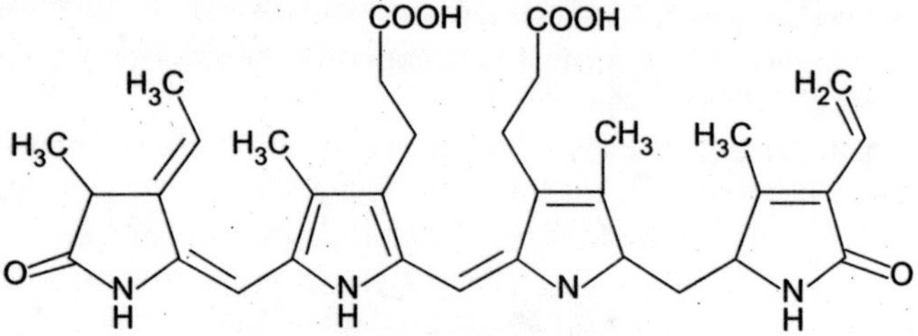

*Fig. 85 : **Phycobilins.***

ANTHOCYANINS

Anthocyanins are water-soluble phytochemicals with a typical red to blue colour. Anthocyanins belong to the group of flavonoids, polyphenolic molecules containing 15 carbon atoms and which can be visualized as two benzene rings joined together with a short three carbon chain. They can be found in tissues of plants, including leaves, stems, roots, flowers and fruits. Anthocyanins occur mainly as glycosides of anthocyanidins such as cyanidin, delphinidin, peonidin, pelargonidin, petunidin and malvidin. Anthocyanins can be found in numerous plants, but high levels are present in acai, backcurrant, blueberry, bilberry, cherry, red grape and purple corn. Anthocyanins have been credited with capacity to modulate cognitive and motor function, to enhance memory. It has a role in preventing age-related declines in neural function. Anthocyanins exerted multiple protective effects against pleurisy. Anthocyanin treatment also downregulated expression of enzymes involved in inflammation in the lung. The antimicrobial activity of anthocyanins in general has been well established. Anthocyanins are powerful antioxidants. Anthocyanins act on cardiovascularsystem. It has anti-cancer activity and anti-inflammatory properties. Anthocyanins prevents arthrosclerosis. Anthocyanins may act as anti-cancer agents by inhibit promotion and progression of tumor cells by stopping the growth of pre-malignant cells, increasing the apoptosis of cancer cells and inhibiting the growth of new blood vessels that nourish tumors. The anti-inflammatory action of anthocyanins may be attributed to its direct and strong antioxidant action but also its regulatory effect on the expression of genes involved in the inflammatory response.

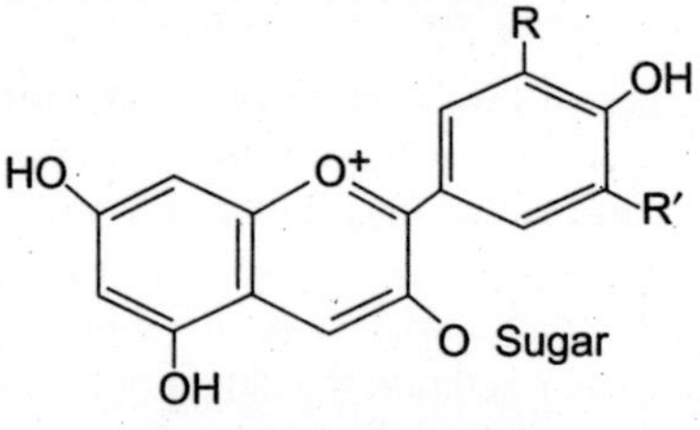

*Fig. 86 : **Anthocyanin.***

CHAPTER 12

SECONDARY METABOLITES

Introduction

Plants produce a variety of compounds that can be divided into primary metabolites and secondary metabolites. Primary metabolites are essential for the survival of the plant and include sugars, proteins and amino acids. Secondary metabolites were once believed to be waste products. They are not essential to the plant's survival, but the plant does suffer without them. Secondary metabolites also have many uses. Some are beneficial, and others can be toxic. Secondary metabolites protect plants against being eaten by herbivores and against being infected by microbial pathogens. They serve as attractants (odour, colour, taste) for pollinators and seed-dispersing animals. They function as agents of plant-plant competition and plant-microbe symbiosis. The ability of plants to compete and survive is therefore profoundly affected by the ecological functions of their secondary metabolites. Plant secondary metabolites can be divided into three chemically distinct groups: **terpenes, phenolics and nitrogen-containing compounds**.

Terpenes

The terpenes or terpenoids are the largest class of secondary metabolites. Most of the diverse substances of this class are insoluble in water. Certain terpenes have well-characterized functions in plant growth or development. The gibberellins, an important group of plant hormones, are diterpenes. Brassinosteroids, another class of plant hormones with growth-regulating functions, originate from triterpenes. The vast majority of terpenes involved in *plant defenses*.

Terpenes are toxins and feeding deterrents to many herbivorous insects and mammals; thus they appear to play important defensive roles in the plant kingdom. For example, monoterpene esters called *pyrethroids*, found in the leaves and flowers of *Chrysanthemum* species, show striking insecticidal activity.

In conifers such as piner, monoterpenes accumulate in resin ducts found in the needles, twigs, and trunk. These compounds are toxic to numerous insects, including bark beetles, which are serious pests of conifer species throughout the world. Many plants contain mixtures of volatile monoterpenes and sesquiterpenes, called essential oils, that lend a characteristic odour to their foliage. Peppermint, lemon, basil, and sage are examples of plants that contain essential oils. The chief monoterpene constituent of lemon oil is limonene; that of peppermint oil is menthol. Essential oils have well-known insect repellent properties.

Phenolic compounds

Plants produce a large variety of secondary compounds that contain a phenol group: a hydroxyl functional group on an aromatic ring. These substances are classified as *phenolic compounds, or phenolics*. Plant phenolics are a chemically heterogeneous group. Some are soluble only in organic solvents, some are water-soluble carboxylic acids and glycosides, and others are insoluble polymers. Many serve as defenses against herbivores and pathogens. Others function of phenolic compounds are in mechanical support, in attracting pollinators and fruit dispersers, in absorbing harmful ultraviolet radiation, or in reducing the growth of nearby competing plants.

The coloured pigments of plants provide visual cues that help to attract pollinators and seed dispersers. These pigments are of two principal types. They are *carotenoids and flavonoids*. Carotenoids are yellow, orange and red terpenoid compounds that also serve as accessory pigments in photosynthesis. The flavonoids also include a wide range of coloured substances. The most widespread group of pigmented flavonoids is the *anthocyanins*, which are responsible for most of the red, pink, purple, and blue colours observed in flowers and fruits. Two other groups of flavonoids found in flowers are *flavones and flavonols*. These flavonoids generally absorb light at shorter wavelengths than do anthocyanins, so they are not visible to the human eye*Isoflavonoids*, which are found mostly in legumes, have several different biological activities.

A second category of plant phenolic polymers with defensive properties is the *tannins*. They are general toxins that can reduce the growth and survival of many herbivores when added to their diets. In addition, tannins act as feeding repellents to a great variety of animals. Mammals such as cattle, deer, and apes characteristically avoid plants or parts of plants with high tannin contents. Plant tannins also serve as defenses against microorganisms.

Nitrogen-containing compounds

A large variety of plant secondary metabolites have nitrogen as part of their structure. *Alkaloids* and *cyanogenic glycosides* are nitrogen containing compounds. Most nitrogenous secondary metabolites are synthesized from common amino acids. The *alkaloids* are a large family of more than 15,000 nitrogen-containing secondary metabolites. They are found in approximately 20% of vascular plant species. As a group, alkaloids are best known for their striking pharmacological effects on vertebrate animals. Alkaloids are usually synthesized from one of a few common amino acids – in particular, lysine, tyrosine, or tryptophan. However, the carbon skeleton of some alkaloids contains a component derived from the terpene pathway. Several different types, including nicotine and its relatives are derived from ornithine, an intermediate in arginine biosynthesis. The B vitamin nicotinic acid (niacin) is a precursor of the pyridine (six-membered) ring of this alkaloid.

Morphine is the first alkaloid isolated from the plant *Papaver sonniferum*, or the opium poppy. It is used as a pain reliever in patients with severe pain levels and cough suppressant. Another example of an alkaloid is **cocaine**. It can be highly dangerous and addictive. However, it has also been used as an anesthetic. Cocaine has long been used by the people of South America to alleviate hunger. Perhaps the most loved and known alkaloid is **caffeine**. It has protective properties for the plants. It is isolated from cocoa, coffee and tea. *Cyanogenic glycosides* release the well-known poisonous gas hydrogen cyanide (HCN). The presence of cyanogenic glycosides deters feeding by insects and other herbivores such as snails and slugs. As with other classes of secondary metabolites, however, some herbivores have adapted to feed on cyanogenic plants and can tolerate large doses of HCN.

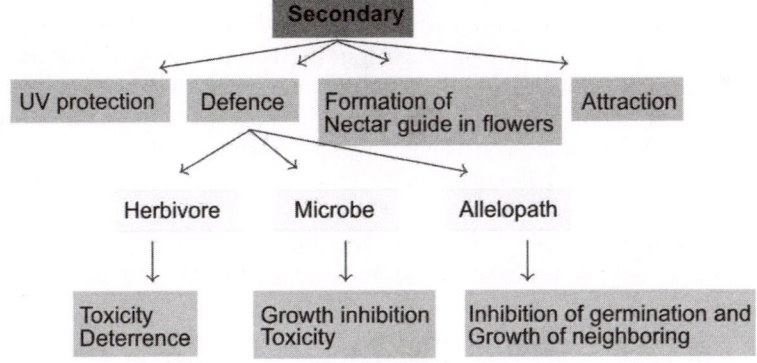

Fig. 87 : Uses of secondary metabolites.

Uses of secondary metabolites

- Induced plant defenses against insect herbivores.
- Jasmonic acid activates many defensive responses.
- Some plant proteins inhibit herbivore digestion.
- Herbivore-induced volatiles have complex ecological functions.
- Plant defenses against pathogens.
- Some antimicrobial compounds are synthesized before pathogen attack.
- Infection induces additional antipathogen defenses.
- Phytoalexins often increase after pathogen attack.
- Some plants recognize specific pathogen-derived substances.
- A single encounter with a pathogen may increase resistance to future attacks.
- Interactions of plants with non-pathogenic bacteria can trigger induced systemic resistance.

Introduction

The term enzyme was first used by Kuhne (1878). Enzymes are also called 'biological catalysts'. Sumner and My Back in 1950 defined the enzymes as "Simple or combined proteins acting as specific catalysts". They affect the life of an organism. to such an extent that life has aptly called as an orderly function of enzymes. Enzymes are chemical catalysts that control various being changed or utilized. The substances on which the enzymes act are called as "Substrates". Enzymes are highly specific in their action (i.e) an enzyme can act on a single or a small group of closely related substrates. During catalytic action, the enzymes do not undergo any permanent modification and regenerated at the end of the reaction.

For example α-amylase acts on starch and produce maltose units. In this reaction α-amylase is the enzyme, starch is the substrate and maltose is the product.

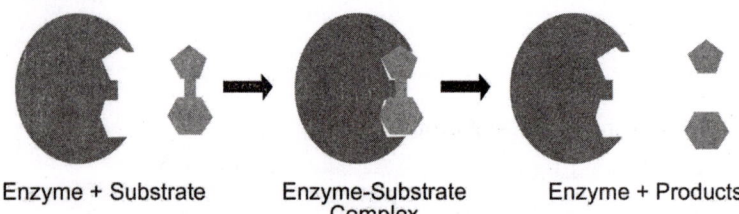

| Enzyme + Substrate | Enzyme-Substrate Complex | Enzyme + Products |

Fig. 88

Characteristics of enzymes

All enzymes are globular proteins.

They increase the rate of reactions.

They are very specific in their reaction.

A single enzyme catalyses a single chemical reaction.

They are sensitive to pH, temperature and substrate concentration.

Some enzyme requires co factor for proper functioning.

Enzymes possess great catalytic power.

Enzymes show varying degree of specificities.

At higher temperatures the rate of the reaction is faster.

The rate of the reaction involving an enzyme is high at the optimum temperature.

Enzymes have an optimum pH range within which the enzymes function is at its peak.

Inorganic substances known as activators increase the activity of the enzyme.

Inhibitors are substances that decrease the activity of the enzyme or inactivate it.

Competitive inhibitors are substances that reversibly bind to the active site of the enzyme, hence blocking the substrate from binding to the enzyme.

Incompetitive inhibitors are substances that bind to any site of the enzyme other than the active site, making the enzyme less active or inactive.

Irreversible inhibitors are substances that from bonds with enzymes making them inactive.

Active site

A restricted region of the enzyme to which the substrate comes and binds and concerns with the process of catalysis is called as the active site. In some enzymes, the active site is a deep groove into which the substrate binds. Specific amino acids are present in the active site which are responsible for the catalytic action. These amino acids are called as 'catalytic' or 'active' amino acids. For example lysozyme has glutamic acid and aspartic acid as catalytic aminoacids. Chymotrypsin which is a proteolytic enzyme contains serine and histidine as catalytic amino acids.

The active site and the other part of the enzyme undergo conformational modification when they come in contact with the substrate

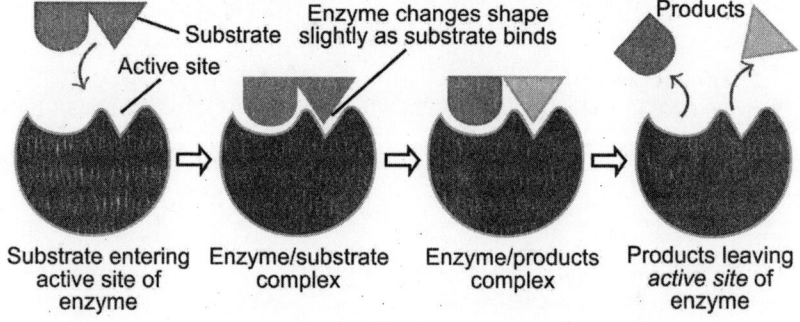

Fig. 89

Koshland's induced fit hypothesis of enzyme-substrate interaction postulates that the active site of the enzyme consists of a number of 'active' contact amino acids which permit the substrate to come close to the reactive groups of the enzyme which there upon undergoes a conformational change, binding the substrate firmly to the enzyme and promoting catalytic activity.

Naming of enzymes

Except the enzymes ptyalin, pepsin, trypsin and renin, all the other enzymes are usually named by adding suffix - **ase** to the main part of the name of the substrate on which they act.

Examples :

Maltase acts on maltose

Lactase acts on lactose

Lipases act on lipids

Proteases act on proteins

Amylases act on starch (amylum).

Classification of Enzymes

The most comprehensive system for the classification of enzymes was devised in 1961 by the Enzyme Commission of International Union of Biochemistry (IUB). The 6 major classes of enzymes are

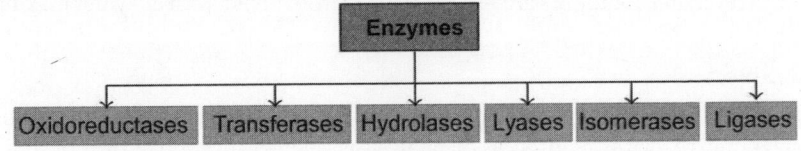

Fig. 90.

1. **Oxidoreductases :** Enzymes catalyzing oxidoreduction reactions between two substrates A and B are called as oxidoreductases

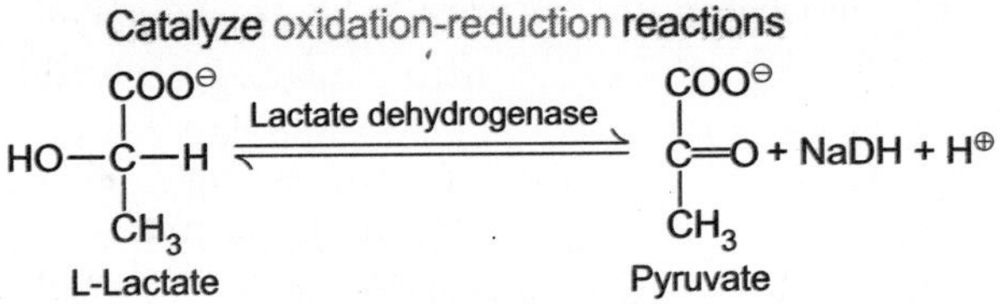

Catalyze oxidation-reduction reactions

$$HO-\underset{\underset{CH_3}{|}}{\overset{\overset{COO^\ominus}{|}}{C}}-H \quad \underset{}{\overset{\text{Lactate dehydrogenase}}{\rightleftharpoons}} \quad \underset{\underset{CH_3}{|}}{\overset{\overset{COO^\ominus}{|}}{C}}=O + NaDH + H^\oplus$$

L-Lactate   Pyruvate

Fig. 91.

2. Transferases : Enzymes catalysing the transfer of a group (x) from one substrate (AX) to another (B) are known as transferases. In this reaction the amino group from alanine is transferred to α-ketoglutarate to form glutamate.

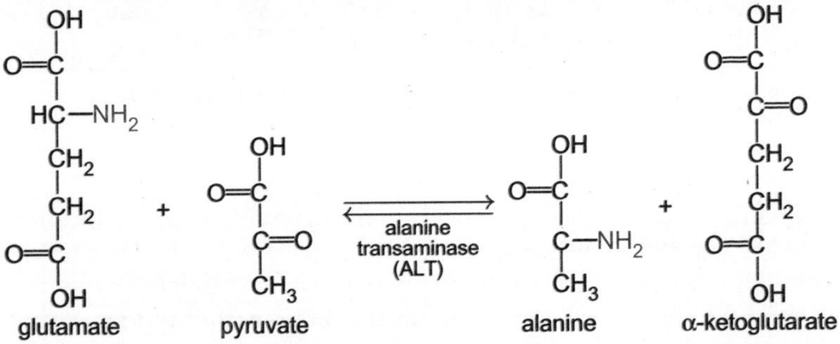

glutamate   pyruvate $\xrightleftharpoons[\substack{\text{alanine} \\ \text{transaminase} \\ \text{(ALT)}}]{}$ alanine   α-ketoglutarate

Fig. 92.

These enzymes are further divided into subclasses on the basis of nature of the group transferred.

Transfer of (a) one carbon compounds

(b) aldehyde or ketonic groups

(c) acyl groups

(d) glycosyl groups

(e) Phosphate groups

(f) Sulphur containing groups

3. Hydrolases : The hydrolases are those enzymes which catalyse hydrolysis reactions i.e the direct addition of water molecule (s) across the bond, which is to be cleaved. The substrate for these enzymes are esters, ethers, peptides and glycosides.

Example : Pepsin. This enzyme is a gastro intestinal enzyme which is proteolytic in nature and involve in the hydrolysis of proteins present in the food.

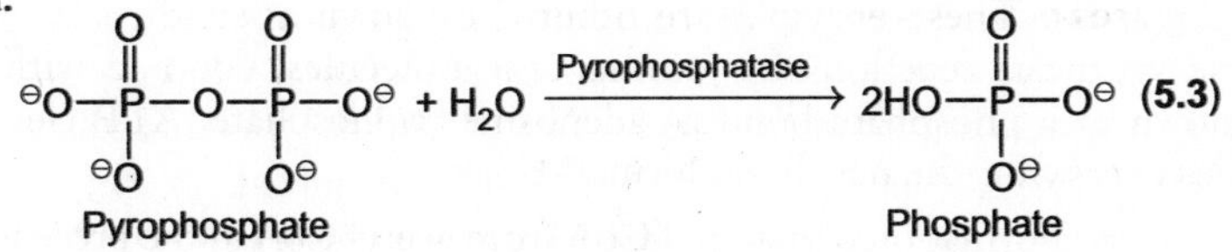

Fig. 93.

The hydrolases are divided into several subclasses, depending on the nature of the group or bond being hydrolysed viz.,

(a) esterases etc. - hydrolyse ester bonds

(b) glycosidases - hydrolyse glycosidic bonds

(c) peptidases - hydrolyse peptide bonds

4. Lyases : The lyases are a smaller class of enzymes that catalyse the removal of a small molecule from a larger substrate molecule. Since the reactions are reversible, lyases may also be considered to catalyse the addition of small molecules to the substrate molecule.

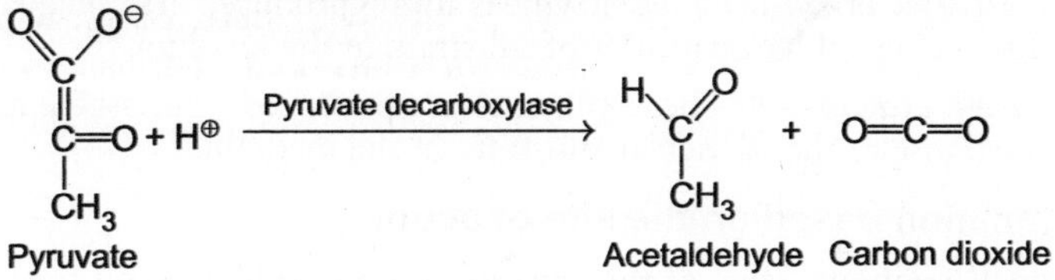

Fig. 94.

5. Isomerases : Isomerases are a general class of **enzymes** which convert a molecule from one **isomer** to another. Isomerases can either facilitate intramolecular rearrangements in which bonds are broken and formed. The general form of such a reaction is as follows:

$$A–B \rightarrow B–A$$

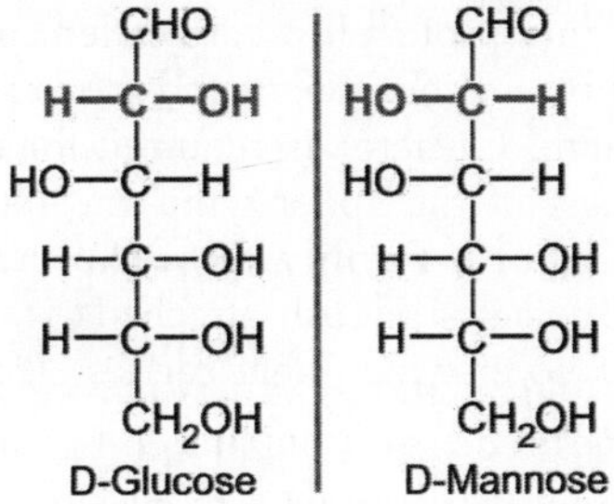

Fig. 95.

Epimers: D-glucose and D-mannose

Isomerases **catalyze** changes within one molecule. They convert one isomer to another, meaning that the end product has the same molecular formula but a different physical structure. **Isomers** themselves exist in many varieties but can generally be classified as **structural isomers** or **stereoisomers**.

6. Ligases : These enzymes are otherwise known as synthetases. They catalyse synthesis reactions by joining two molecules, coupled with the breakdown of a phosphate bond of adenosine triphosphate. ATP cleavage provides energy for the new bond formation.

Example : Formation of malonyl CoA from acetyl CoA in the presence of acetyl CoA carboxylase.

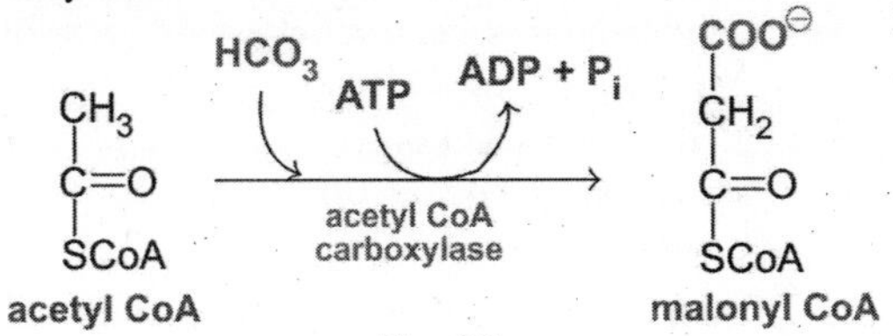

Fig. 96.

Classification based on the substrate concentration

Constitutive enzyme : This enzyme is always produced in small quantities in the absence or milder quantities of substrate in the medium.

Adaptive enzymes : These enzymes are produced in unusable amount only in response to the particular substrate in the medium.

Classification based on the site of action

Endo Enzymes : Most of the enzymes produced by the cell for the cell function within the cell are called endo enzyme. It is also called as intracellular enzymes. Eg: Helicases.

Exo Enzymes : Some enzymes are liberated by the living cells and catalyse reactions in the cells environment. Such enzymes are called exo exymes. It is also called as extracellular enzyme. Eg: Amlyase.

Chemical Nature of Enzymes

Many enzymes are pure proteins and are called apoenzyme. Some of the enzyme also have non protein component and are called a cofactor. Eg. iron, zinc, magnesium, calcium. Cofactor is required for catalytic activity. The complete enzyme consist of the apoenzyme and its cofactor is called the holoenzyme. If the co factor is firmly attached to the apoenzyme are called prosthetic group. If the cofactor is loosly attached to the apoenzyme are called co enzyme. NAD is a co enzyme. That carries electrons. Metal ions may bound to apo enzyme and act as a cofactor.

For example, pyruvate decarboxylase is an enzyme which catalyses the decarboxylation of pyruvate to form acetaldehyde. The non-protein part of the enzyme is thiamine pyro phosphate without which the reaction can not be proceeded.

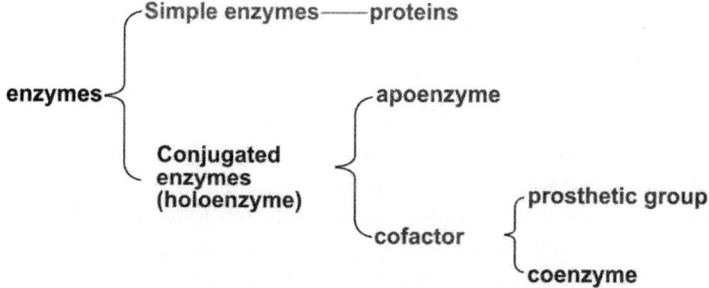

Holoenzyme = Apoenzyme + Cofactors

Fig. 97 : Holo enzyme.

Structure of Enzymes

Enzymes are proteins, like the proteins the enzymes contain chains of amino acids linked together. The characteristic of an enzyme is determined by the sequence of amino acid arrangement. When the bonds between the amino acid are weak, they may be broken by conditions of high temperatures or high levels of acids. When these bonds are broken, the enzymes become nonfunctional. The enzymes that take part in the chemical reaction do not undergo permanent changes and hence they remain unchanged to the end of the reaction.

Enzymes are highly selective, they catalyze specific reactions only. Enzymes have a part of a molecule where it just has the shape where only certain kind of substrate can bind to it, this site of activity is known as the *'active site'*. The molecules that react and bind to the enzyme is known as the *'substrate'*.

Most of the enzymes consists of the protein and the non protein part called the *'cofactor'*. The proteins in the enzymes are usually globular proteins. The protein part of the enzymes are known *'apoenzyme'*, while the non-protein part is known as the cofactor. Together the apoenzyme and cofactors are known as the *'holoenzyme'*.

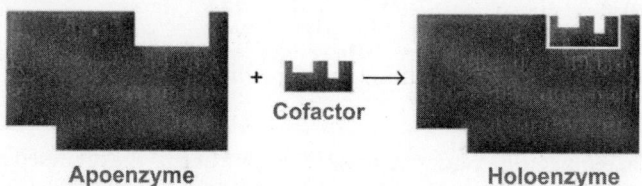

Fig. 98 : Enzyme Nature.

Cofactors may be of three types: prosthetic groups, activators and coenzymes.

Prosthetic groups are organic groups that are permanently bound to the enzyme.

Example : Heme groups of cytochromes and bitotin group of acetyl-CoA carboxylase.

Activators are cations- they are positively charged metal ions.

Example : Fe - cytochrome oxidase, CU - catalase, Zn - alcohol dehydrogenase, Mg - glucose - 6 - phosphate, etc.

Coenzymes are organic molecules, usually vitamins or made from vitamins. they are not bound permanently to the enzyme, but they combine with the enzyme-substrate complex temporarily.

Example : FAD - Flavin Adenine Dinucleotide, FMN - Flavin Mono Nucleotide, NAD - Nicotinamide Adenine Dinucleotide, NADP - Nicotinamide Adenine Dinucleotide Phosphate.

Factors influencing enzyme activity

The activity of enzymes is markedly affected by several factors. These factors are

1. pH

2. temperature

3. substrate concentration

4. metal ions (activators)

5. inhibitors

6. enzyme concentration etc.

pH

All the enzymes have a particular pH at which their activity is maximal; above or below this pH the activity is low. The pH at which the enzyme shows maximum activity is known as optimum pH. Some of the enzymes and their optimum pH are

(a) Pepsin - 2.0

(b) Urease - 7.0

(c) Salivary amylase - 6.8

(d) Alkaline phosphatase - 9.9

Only in this optimum pH, ionisation of active amino acids in enzymes and substrate are favoured for ES complex formation.

Temperature

Rise in temperature causes increase in the rate of enzyme catalysed reactions up to a certain temperature i.e. about 45°C. Above which the activity declines due to denaturation of enzymes (due to their protein nature). As the enzyme is denatured and inactivated, the reaction which it catalyses slows down and ultimately stops. So the temperature at which the enzyme shows maximum activity is known as optimum temperature. The optimum temperature of most of the enzymes is found to be 37°C.

Substrate concentration

The enzyme action rate is proportional to the concentration of substrate. But this is true upto a certain concentration after which the increase in concentration of substrate does not further increase the velocity of the reaction. Since the number of active sites on an enzyme molecule are limited, a stage will come when all of them have filled with the substrate molecules. This is known as saturation of enzyme. Now, since none of the active sites of the enzyme is free, further addition of the substrate molecule will not increase the product formation. It was Michaelis and Menten in 1913, who proposed a successful explanation for the effect of substrate concentrtaion on the enzyme activity. According to them the enzyme 'E', and the substrate 'S' combine rapidly to form a complex, the enzyme substrate complex 'ES'. The complex then breaks down relatively, slowly to form the product of the reaction. The enzyme regenerated can involve in another round of catalysis.

Effect of activators

Divalent ions, like Mg^{2+}, Cu^{2+}, Mn^{2+}, Zn^{2+} and monovalent ions such as Na^+ and K^+ are required for the activity of many enzymes. For example, amylases need Cl^- ions, Zn^{2+} ions for carbonic anhydrase action, Fe^{2+} and Cu^{2+} ions are required for enzymes involved in redox reactions. Several peptidases are activated by Mn^{2+}, Zn^{2+} or Co^{2+}. Enzymes requiring metal ions or enzymes which contain metal ions in their structure are called as metallo enzymes.

Effect of concentration of enzyme

The velocity of an enzymatic reaction is directly proportional to the concentration of enzyme. In case the enzyme concentration is doubled then as much as twice active site become available to combine with the substrate, provided an excess of substrate is present and so the maximum velocity is also doubled. At a fixed concentration of the substrate a level is reached when all the substrate molecules are utilised and no more change in velocity of the reaction takes place.

Enzyme inhibition and types

It is a process of inhibiting the catalytic activities of enzyme using inhibitors. In other words This phenomenon in which the enzyme activity is decreased by the presence of inhibitors is known as enzyme inhibition. There are three types of enzyme inhibition. They are

(a) Reversible inhibition

(b) Irreversible inhibition

(c) allosteric inhibition.

Inhibitors

Chemical substances which reduce the activity of enzymes are called as inhibitors. They may be small inorganic ions such as cyanide which inhibits the enzyme cytochrome oxidase or much more complex molecules such as diisopropyl phospho fluoridate which inhibit acetyl choline esterase.

Competitive inhibition

It is a type of Reversible inhibition. In this inhibitors attach to enzymes with non-covalent interactions such as **hydrogen bonds, hydrophobic interactions** and **ionic bonds**. Multiple weak bonds between the inhibitor and the active site combine to produce strong and specific binding. The inhibition can be reversed if the inhibitor is removed.

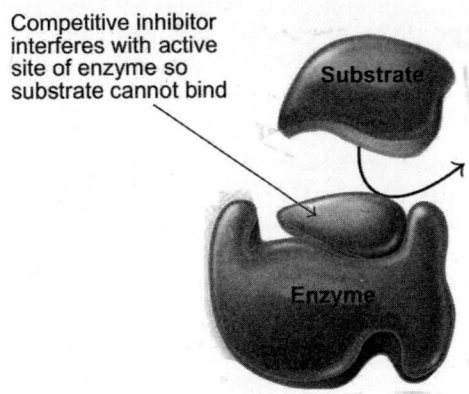

Competitive inhibitor interferes with active site of enzyme so substrate cannot bind

Substrate

Enzyme

Fig. 99.

This type of inhibition occurs when the structure of inhibitor resembles that of the substrate. The inhibitor competes with the proper substrate for binding at the active site of the enzyme. In this type of inhibition, both ES complex and EI complex (enzyme – inhibitor complex) are formed during the reaction. The relative amounts of the two complexes depend partly upon the affinity of the enzyme towards the substrate and inhibitor and partly upon the relative concentration of substrate and the inhibitor. Thus if the inhibitor is present in sufficiently high concentration, it can displace the substrate entirely and thus blocks the reaction completely). The enzyme succinate dehydrogenease is a example for competitive inhibition with succinic acid as a substrate. Malonic acid, Glutaric acid and oxalic acid are structurally similar compete each other for bing active site of succinate dehydrogenease. Ethanol and methanol are similarly binds on the active site of alcohol dehydrogenase. Antimetabolite and antivitamins are works like a competitive inhibitors. Antimetabolites are the structural analogs of substrates and thus considered as a competitive inhibitors.

Uncompetitive inhibition

In this type of inhibition, the inhibitor combines with enzyme - substrate complex to give an inactive enzyme - substrate - inhibitor complex which cannot undergo further reaction to yield the product. In this type, the degree of inhibition may increase when the substrate concentration is increased. This inhibition cannot be reversed by increasing the concentration of substrate. So it is also called irreversible enzyme inhibition. In this type of inhibition, inhibitors does not bind on the active site of enzyme. Many enzymes like pepsin, urease and succinic dehydrogense have SH group. The compound iodoacetate combines with SH group and inactivates the enzyme. Heavy metals able to inhibit enzymes by the method only (**Fig.100**).

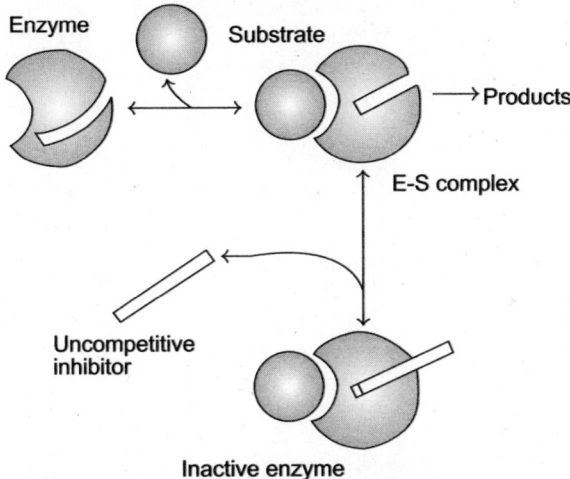

Fig. 100 : **Mode of action of uncompetitive inhibitor.**

Examples :

(a) Effect of iodoacetamide on –SH group containing enzymes

(b) Effect of diisopropyl phosphofluoridate on acetyl choline esterase. These two inhibitors completely inactivate the respective enzymes. This inhibition can be partially reversible.

Allosteric inhibition

It is a reversible enzyme inhibition. Inhibitor binds otherthan active site of an enzyme. This type of inhibition is otherwise known as end product inhibition. The inhibitor binds with the modulator binding site (or) allosteric site (i.e. other site) of the enzyme. The inhibitor present at the allosteric site may affect the conformation at the active site with the result it becomes difficult for the enzyme to take up the substrate molecule, and in the extreme case, the enzyme completely fails to take up the substrate molecule (**Fig.101**).

Example : when isoleucine production increases, as a regulatory mechanism, it binds with threonine deaminase in the allosteric site and inhibit further binding of the substrate with the enzyme and ultimately production of isoleucine is stopped. This inhibition is otherwise known as feed back inhibition. Many metabolic reactions in our body are regulated by means of allosteric enzymes.

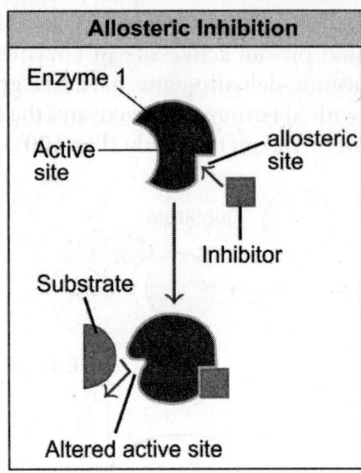

Fig. 101.

As the end product prevents the catalyzing mechanism of the same reaction, the inhibition is also called feed back inhibition.

Iso enzymes

Certain enzymes may exist in two or more forms which have the same catalytic activity but are distinct physically, chemically and electrophoretically. These forms exist in different tissues.

Example :

(a) Lactate dehydrogenase exist in 5 different iso enzymic forms LD_1, LD_2, LD_3, LD_4, LD_5 and perform the same function of the conversion of lactate to pyruvate. LD_1 is predominant in heart and LD_5 in muscle and liver.

(b) Creatine kinase exist in 3 different iso enzymic forms **BB, MM and MB**. These isoenzymes perform the same function of conversion of creatine to creatine phosphate. BB form exist in brain, MM form in muscle and MB form in heart.

Mechanism of enzyme action

The breaking of substrate into end products by an enzyme is called enzyme action. Enzyme is active in catalytic action of biochemical reaction. They act on substrate and forms a complex after interactions with the enzyme is called active center. The enzyme and substrate forms a complex at the active centre.

This binding action makes both enzyme and substrate stable. The interaction between substrate and enzyme may be either ionic bonds and hydrogen bonds or Van der Waal forces. The active sites of enzyme have some special groups such as NH_2 COOH, –SH etc. which bind the substrate though above bonds to form a transitional (intermediate) compound called enzyme-substrate complex (ES). This reaction is exergonic and releases some energy which raises energy level of the substrate molecule.

$$E \; + \; S \; \longleftrightarrow \; \text{E.S Complex}$$
$$\text{(Enzyme)} \quad \text{(Substrate)}$$

Types of Mechanisms of Enzymes : There are two types of mechanisms involved to explain substrate-enzyme complex formation; lock and key theory (template model), and induced-fit theory.

(i) Lock and Key Theory : This is a most accepted theory for enzyme action This theory is first postulated in 1894 by Emil Fischer. This theory postulated that, the lock is the enzyme and the key is the substrate. Only the correctly sized key (substrate) fits into the key hole (active site) of the lock (enzyme). Smaller keys, larger keys, or incorrectly positioned teeth on keys (incorrectly shaped or sized substrate molecules) do not fit into the lock (enzyme). Only the correctly shaped key opens a particular lock.

Enzyme is analogous to key, where the geometrical configuration of socket is fixed. Similarly substrate has also got fixed geometrical configuration like that of key. A particular lock can be opened or closed by a particular key. According to the particular substrate can be found at active site of particular enzyme forming substrate-enzyme complex.

Enzyme-substrate complex remains in tight fitting and active sites of enzymes are complementary to substrate molecules. Subsequently, enzyme-substrate complexes result in the transformation of substrate into the product formation due to activity of reaction sites.

Since product has lower free energy, it is released. Enzymes are fixed to receive another molecule of substrate and thus enzyme activity continues. In this analogy, the lock is the substrate and the key is the enzyme. Only the correctly sized key (substrate) fits into the key hole (active site) of the lock (enzyme). Smaller keys, larger keys, or incorrectly positioned teeth on keys (incorrectly shaped or sized substrate molecules) do not fit into the lock (enzyme) (**Fig.102**).

(ii) Induced Fit Theory : In 1958, Koshland modified the Fischer's model for the formation of an enzyme-substrate complex to explain the enzyme property more efficiently. According the Fischer's model the nature of the active site of enzyme is rigid, but it is able to be pre-shaped to fit the substrate.

Koshland explains that the enzyme molecule does not retain its original shape and structure, but the contact of the substrate induces some geometrical changes in the active site of the enzyme molecule. The enzyme molecule is made to fit completely the configuration and active centers of the substrate. At the same time, other amino acid residues may become buried in the interior of the molecule.

Lock and Key Analogy

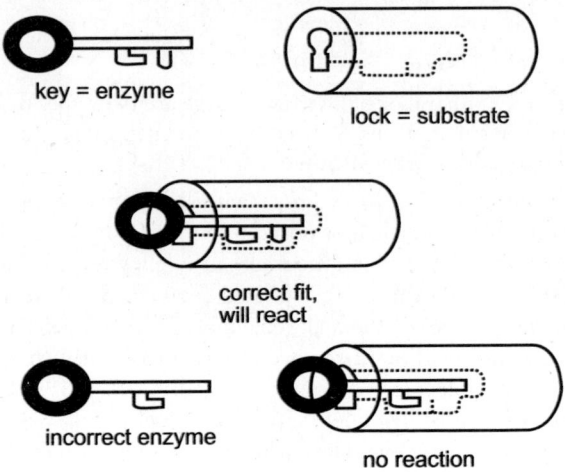

Fig. 102 : **Lock and Key model.**

The hydrophobic and charged group both are involved in substrate binding. A phosphoserine (-P) and SH group of cysteine residue are involved in catalysis.

Residue of the other amino acid such as lysine (Lys) and methionine (Met) are not involved in either binding or catalysis. In the absence of substrate, the substrate binding group and catalytic group are far apart from each other.

But the contact of the substrate induces a conformational changes in the enzyme molecule and aligns both the groups for substrate binding and catalysis. Simultaneously, the spatial orientation of the other region also changed. This causes the lysine and methionine much closer (**Fig.103**).

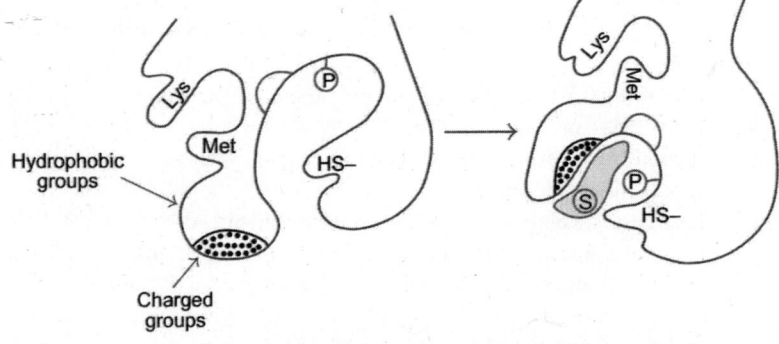

Fig. 103 : **Conformational changes brought about by induced fit in an enzyme molecule.**

Enzyme Kinetics : A number of approaches are now available to study the mechanism of enzyme action including knowledge of complete 3-D structure, site directed mutagenesis and protein engineering. Enzyme kinetics follows the principles of general chemical reaction kinetics; however, show a distinctive feature of saturating (**Fig.104**). At lower substrate concentration, the initial reaction velocity is proportional to substrate concentration (1st order reaction). Further increase in substrate concentration does not affect the reaction rate and the latter became constant (zero order reaction).

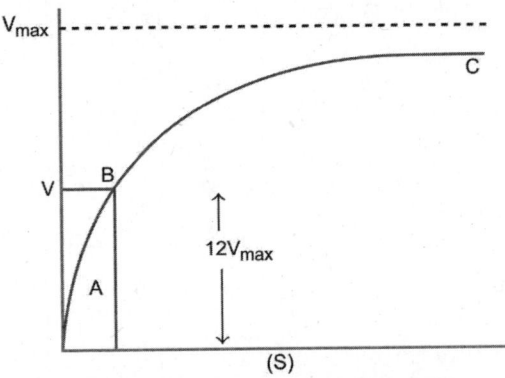

Fig. 104 : **Kinetics of enzyme-substrate relationship (Michaelis-Menten hypothesis).**

In hypothetical one substrate reaction Michaelis-Menten theory, enzymes first combine with substrate to from enzyme-substrate (ES) complex.

This ES complex then breaks in second step to release free enzyme and product as:

$$E + S \longleftrightarrow ES \qquad (1)$$

$$ES \longleftrightarrow E + P \qquad (2)$$

According to above equations, initial velocity of complete reaction equals the breakdown of enzyme substrate complex. Hence,

Vo = Kb (ES) (3)

Where, Vo = initial velocity,

(ES) = concentration of enzyme substrate complex

However, neither of the two parameters can be determined directly, and an alternative expression of Vo is required. This can be done by considering 23″ order rate equation for formation of ES from E and S.

$$d(ES)/dt = Ka(E) (S) = Ka[ET] - (ES)] (S) \qquad (4)$$

Where, Ka= second order rate constant

(ET) = total enzyme concentration

(ES) = concentration of enzyme substrate complex

Since, starting of reaction is being considered, formation of (ES) by reaction (2) may be neglected because in the beginning for reaction in forward direction, when (S) is high and (P) is zero.

Rate equation for degradation of ES can be expressed as sum of two reactions-first reaction yielding the product and second reaction yielding E + S. Then,

$$-d(ES)/dt = Ka \ (ES) + Kb(ES) \qquad (5)$$

However, in the steady state, when rate of formation of ES is equal to its breakdown, then equation 4 = 5

Thus:

$$(ES)/dt = -d(ES)/dt \ (6)$$

$$or, Ka \ [ET]-[ES] \ (S) = Ka \ (ES) + Kb \ (ES)$$

$$or, [S] \ [(ET)-(ES)]/(ES) = Ka + Kb/Ka = Km \qquad (7)$$

(Ka+Kb/Ka), i.e. expressed as Km is known as Michaehs-Menten constant. Rearranging the equation 7 gives,

$$(ES) = (ET) \ (S)/Km + (S) \qquad (8)$$

The value of (ES), when expressed in equation :

$$Vo = Kb \ (ES)$$

$$Vo = Kb \ (ET) \ (S)/ \ Km + (S) \ (9)$$

When substrate concentration is high, all enzymes in system are present as ES complex.

Hence enzyme will be saturated, and reach the maximum velocity (V max), given as:

$$Vmax = Kb \ (ET)$$

Putting this value in equation 9, we get

$$V_0 = Vmax \ (S)/Km + (S) \qquad (10)$$

This is Michaelis-Menten equation i.e. the rate equation for a one substrate enzyme catalyzed reaction considering special case when initial reaction rate is exactly half of Vmax, then by equation 10

$$Vmax/2 = Vmax \ (S)/Km + (s)$$

$$or, Km + (S) = 2(S)$$

$$or, Km = (S)$$

Thus Michaelis-Menten constant is equal to substrate concentration at which initial reaction velocity is half of maximum velocity.

Km varies according to substrate, pH, and temperature and is not a fixed value.

Reciprocating the equation 10, we get

$$1/V_0 = Km + (S)/Vmax \ (S)$$

$$or, \ W_0 = Km/Vmax \ (S) \ 4- \ 1/Vmax \qquad (11)$$

It is Lineweaver-Burk equation, which gives a straight line when $1/V_0$ is plotted against 1/S (**Fig.105**).

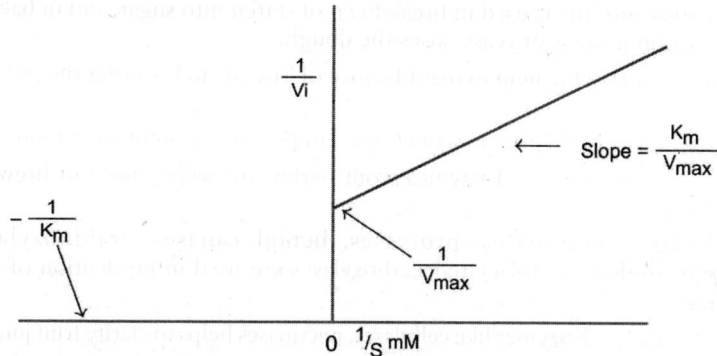

Fig. 105 : *Enzyme-substrate relationship (uneweover-Burk Plot).*

Importance of enzymes

1. Enzymes catalyse many biological reactions and enhance the rate of product formation in metabolic pathways.

2. Some enzymes in blood are used as diagnostic indicators of various diseases. For example the level of transaminases are elevated in blood during jaundice - a liver disorder.

3. Some enzymes are used for therapeutic purposes

      (a) Penicillinase - to treat patients allergic to penecillin

      (b) Asparaginase - to treat leukemia

      (c) Diastase - to treat indigestion

Biological Functions of Enzymes

Enzymes perform a wide variety of functions in living organisms.

They are major components in signal transduction and cell regulation, kinases and phosphatases help in this function.

They take part in movement with the help of the protein myosin which aids in muscle contraction.

Also other ATPases in the cell membrane acts as ion pumps in active transport mechanism.

Enzymes present in the viruses are for infecting cell.

Enzymes play a important role in the digestive activity of the enzymes.

Amylases and proteases are enzyme that breakdown large molecules into absorbable molecules.

Variuos enzymes work together in a order forming metabolic pathways. Example: Glycolysis.

Industrial Application of Enzymes

Food Processing - Amylases enzymes from fungi and plants are used in production of sugars from starch in making corn-syrup.

Catalyze enzyme is used in breakdown of starch into sugar, and in baking fermentation process of yeast raises the dough.

Proteases enzyme help in manufacture of biscuits in lowering the protein level.

Baby foods - Trypsin enzyme is used in pre-digestion of baby foods.

Brewing industry - Enzymes from barley are widely used in brewing industries.

Amylases, glucanases, proteases, betaglucanases, arabinoxylases, amyloglucosidase, acetolactatedecarboxylases are used in prodcution of beer industries.

Fruit juices - Enzymes like cellulases, pectinases helps to clarify fruit juices.

Dairy Industry - Renin is used inmanufacture of cheese. Lipases are used in ripening blue-mold cheese. Lactases breaks down lactose to glucose and galactose.

Meat Tenderizes - Papain is used to soften meat.

Starch Industry - Amylases, amyloglucosidases and glycoamylases converts starch into glucose and syrups.

Glucose isomerases - production enhanced sweetening properties and lowering calorific values.

Paper industry - Enzymes like amylases, xylanases, cellulases and liginases lower the viscosity, and removes lignin to soften paper.

Biofuel Industry - Enzymes like cellulases are used in breakdown of cellulose into sugars which can be fermented.

Biological detergent - proteases, amylases, lipases, cellulases, asist in removal of protein stains, oily stains and acts as fabric conditioners.

Rubber Industry - Catalase enzyme converts latex into foam rubber.

Molecular Biology - Restriction enzymes, DNA ligase and polymerases are used in genetic engineering, pharmacology, agriculture, medicine, PCR techniques, and are also important in forensic science.

CARBOHYDRATE METABOLISM

GLYCOLYSIS

Oxidation of glucose to pyruvate is called glycolysis. It was first described by Embden-Meyerhof and Parnas. Hence it is also called as Embden-Meyerhof pathway.

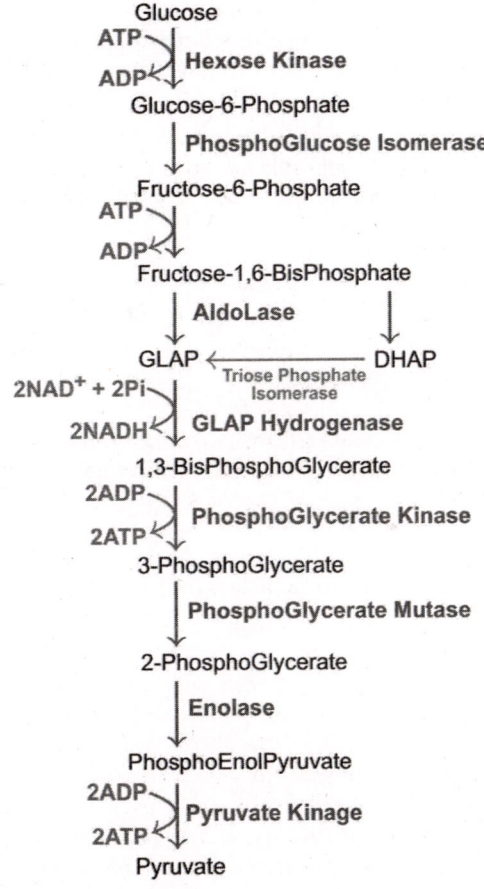

FIG. 106 : *Glycolysis.*

Reactions of glycolytic pathway

Series of reactions of glycolytic pathway which degrades glucose to pyruvate are represented below. The sequence of reactions occurring in glycolysis may be considered under four stages.

Stage I

This is a *preparatory phase*. Before the glucose molecule can be split, the rather asymmetric glucose molecule is converted to almost symmetrical form, fructose 1,6-diphosphate by donation of two phosphate groups from ATP.

The glycolytic pathway

1. **Uptake of glucose by cells and its phosphorylation :** Glucose is freely permeable to cells. Glucose enters through active' transport. Glucose is phosphorylated to form glucose 6-phosphate. The enzyme involved in this reaction is glucokinase. This reaction is irreversible.

2. **Conversion of glucose 6-phosphate to fructose 6-phosphate :** Glucose 6-phosphate is converted to fructose 6-phosphate by the enzyme phosphogluco isomerase.

3. **Conversion of fructose 6-phosphate to fructose 1,6 diphosphate :** Fructose 6-phosphate is phosphorylated irreversibly at 1 position catalyzed by the enzyme phosphofructokinase to produce fructose 1,6-diphosphate.

Stage II

1. **Actual splitting of fructose 1,6 diphosphate :** Fructose 1,6 diphosphate is split by the enzyme aldolase into two molecules of triose phosphates, an aldotriose-glyceraldehyde 3-phosphate and one ketotriose - dihydroxy acetone phosphate. The reaction is reversible. There is neither expenditure of energy nor formation of ATP.

2. **Interconvertion of triose phosphates :** Dihydroxy acetone phosphate is converted to Glyceraldehyde 3-phosphate using an enzyme called Triose phosphate isomerase

Stage III

It is the energy yielding stage. Reactions of this type in which an aldehyde group is oxidised to an acid are accompanied by liberation of large amounts of potentially useful energy.

1. **Oxidation of glyceraldehyde 3-phosphate to 1,3-bisphospho-glycerate :** Glycolysis proceeds by the oxidation of glyceraldehydes

3-phosphate to form 1,3-bisphosphoglycerate. The reaction is catalyzed by the enzyme glyceraldehyde 3-phosphate

2. **Conversion of 1, 3-bisphosphoglycerate to 3-phosphoglycerate :** The reaction is catalyzed by the enzyme phosphoglycerate kinase. The high energy phosphate bond at position - 1 is transferred to ADP to form ATP molecule.

Stage IV

It is the recovery of the phosphate group from 3-phosphoglycerate. The two molecules of 3-phosphoglycerate, the end-product of the previous stage, still retains the phosphate group, originally derived from ATP in *Stage I.*

1. **Conversion of 3-phosphoglycerate to 2-phosphoglycerate** : 3-phosphoglycerate formed by the above reaction is converted to 2-phosphoglycerate, catalyzed by the enzyme phosphoglycerate mutase.

2. **Conversion of 2-phosphoglycerate to phosphoenol pyruvate** : The reaction is catalyzed by the enzyme enolase, the enzyme requires the presence of either Mg^{2+} or Mn^{2+} ions for activity.

3. **Conversion of phosphoenol pyruvate to pyruvate** : Phosphoenol pyruvate is converted to pyruvate, the reaction is catalysed by the enzyme pyruvate kinase. The high energy phosphate group of phosphoenol pyruvate is directly transferred to ADP, producing ATP. The reaction is irreversible.

Summary of glycolysis

During glycolysis NAD^+ is reduced to NADH. At the same time, glyceraldehyde 3-phosphate is oxidized to 1,3-bisphosphoglycerate. To conserve the coenzyme NAD^+, NADH must be reoxidized. Under anaerobic conditions this is done when pyruvic acid is converted to lactic acid. In the presence of oxygen, NADH, can be oxidized to NAD^+ with the help of the respiratory enzymes.

Net gain = 8 ATP

Reactions Catalyzed	ATP used	ATP formed
Stage I		
1. Glucokinase (for phosphorylation)	1	
2. Phosphofructokinase I (for phosphorylation)	1	
Stage II		
3. Glyceraldehyde 3-phosphate dehydrogenase (oxidation of 2 NADH in respiratory chain)		6
4. Phosphoglycerate kinase (substrate level phosphorylation)		2
Stage IV		
5. Pyruvate kinase (substrate level phosphorylation)		2
Total :	2	10

TRICARBOXYLIC ACID CYCLE (TCA CYCLE)

This cycle is the aerobic phase of carbohydrate metabolism and follows the anaerobic pathway from the stage of pyruvate and is called as citric acid cycle or TCA cycle. The name citric acid cycle stems from citric acid which is formed in the first step of this cycle. This cycle is also named "Kerbs cycle" after H.A. Krebs, an English biochemist who worked on it.

Under aerobic conditions, pyruvate is oxidatively decarboxylated to acetyl coenzyme A (active acetate) before entering the citric acid cycle.

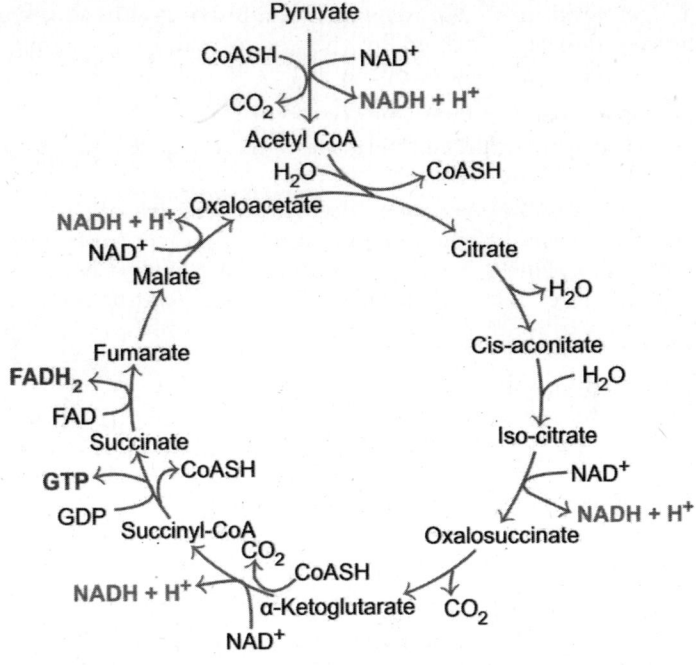

*FIG. 107 : **TCA Cycle.***

1. Formation of citrate : The first reaction of the cycle is the condensation of acetyl CoA with oxaloacetate to form citrate, catalyzed by citrate synthase. This is an irreversible reaction.

2. Formation of isocitrate via cis aconitate : The enzyme aconitase catalyzes the reversible transformation of citrate to isocitrate, through the intermediary formation of cis aconitate.

3. Oxidation of isocitrate to α-ketoglutarate and CO_2 : In the next step, isocitrate dehydrogenase catalyzes oxidative decarboxylation of isocitrate to form α-ketoglutarate.

4. Oxidation of α-ketoglutarate to succinyl CoA and CO_2 : The next step is another oxidative decarboxylation, in which α-ketoglutarate is converted to succinyl CoA and CO_2 by the action of the α-ketoglutarate dehydrogenase complex. The reaction is irreversible.

5. Conversion of succinyl CoA to succinate : The product of the preceding step, succinyl CoA is converted to succinate to continue the cycle. GTP is formed in this step (substrate level phosphorylation). The enzyme that catalyzes this reversible reaction is called succinyl CoA synthetase or succinic thiokinase.

6. Oxidation of succinate to fumarate : The succinate formed from succinyl CoA is oxidized to fumarate by the enzyme succinate dehydrogenase.

7. Hydration of fumarate to malate : The reversible hydration of fumarate to malate is catalyzed by fumarase.

8. Oxidation of malate to oxaloacetate : The last reaction of the citric acid cycle is, NAD linked malate - dehydrogenase which catalyses the oxidation of malate to oxaloacetate.

Energy yield from TCA cycle

If one molecule of the substrate is oxidized through NADH in the electron transport chain three molecules of ATP will be formed and through FADH2, two ATP molecules will be generated.

As one molecule of glucose gives rise to two molecules of pyruvate by glycolysis, intermediates of citric acid cycle also result as two molecules.

Table **3.2**

Reactions	No. of ATP formed
1. 2 isocitrate 2 α-ketoglutarate	
(2NADH + 2H+) (2X_3)	6
2. 2 α-ketoglutarate 2 succinyl CoA	
(2NADH + 2H+) (2'3)	6
3. 2 succinyl CoA 2 succinate	
(2GTP = 2ATP)	2
4. 2 succinate 2 Fumarate	
(2FADH_2) (2'2)	4
5. 2 malate 2 oxaloacetate	
(2NADH + 2H+) (2'3)	6
Total No. of ATP formed	24

HMP SHUNT PATHWAY

Glycolysis and citric acid cycle are the common pathways. A number of alternative pathways are also discovered. The most important one is Hexose Monophosphate Shunt Pathway (HMP shunt). The pathway occurs in the extra mitochondrial soluble portion of the cells.

HMP shunt generates a different type of metabolic energy - the reducing power. Some of the electrons and hydrogen atoms of fuel molecules are conserved for biosynthetic purposes rather than ATP formation. This reducing power of cells is NADPH (reduced nicotinamide adenine dinucleotide phosphate).

The fundamental difference between NADPH and NADH (reduced nicotinamide adenine dinucleotide) is that NADH is oxidised by the respiratory chain to generate ATP whereas NADPH serves as a hydrogen and electron donor in reductive biosynthesis, for example in the biosynthesis of fatty acids and steroids. The first reaction of the pentose phosphate pathway is the dehydrogenation of glucose 6-phosphate by glucose 6-phosphate dehydrogenase to form 6-phosphoglucono d-lactone.

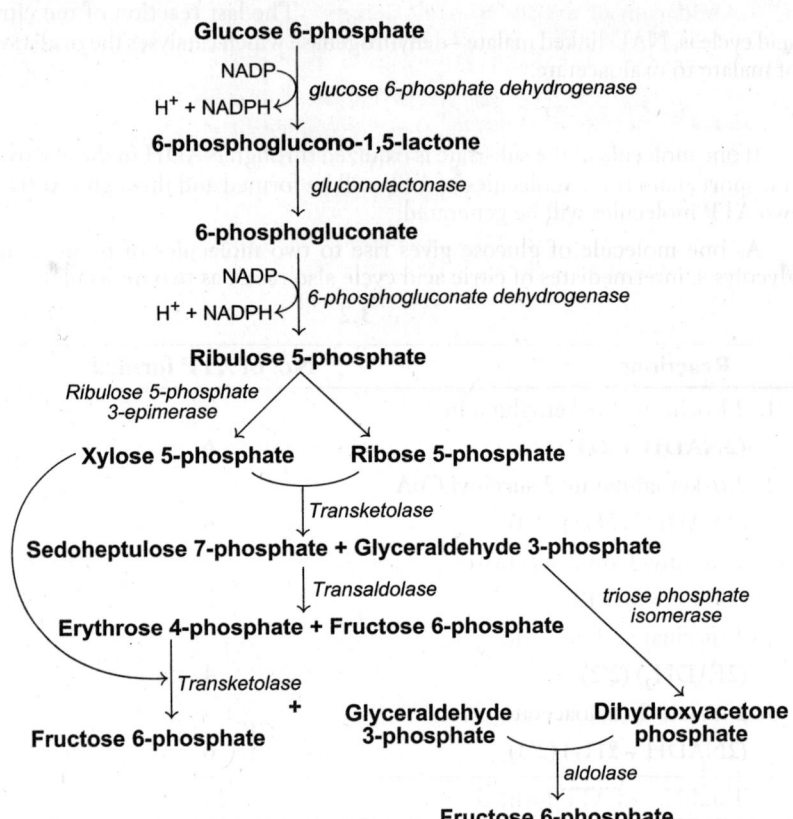

Fig. 108 : HMP Pathway.

Step 1

Glucose 6-phosphate in the presence of NADP and the enzyme glucose 6-phosphate dehydrogenase, forms 6-phospho glucono-d-lactone. The first molecule of NADPH is produced in this step.

Step 2

The 6-phospho glucono d-lactone is unstable and the ester spontaneously hydrolyses to 6-phosphogluconate. The enzyme that catalyses the reaction is lactonase

Step 3

6-phospho gluconate further undergoes dehydrogenation and decarboxylation by 6-phosphogluconate dehydrogenase to form the ketopentose, D-ribulose 5-phosphate. This reaction generates the second molecule of NADPH.

Step 4

The enzyme phosphopentose isomerase converts ribulose 5-phosphate to its aldose isomer, D-ribose 5-phosphate. In some tissues, the hexose phosphate pathway ends at this point.

The net result is the production of NADPH, a reductant for biosynthetic reactions, and ribose 5-phosphate, a precursor for nucleotide synthesis.

ED PATHWAY

The **Entner-Doudoroff pathway** begins with the same reactions as the pentose phosphate pathway, the formation of glucose 6-phosphate and 6-phosphogluconate. Instead of being further oxidized, 6-phosphogluconate is dehydrated to form 2-keto-3-deoxy-6-phosphogluconate or KDPG, the key intermediate in this pathway. KDPG is then cleaved by KDPG aldolase to pyruvate and glyceraldehydes 3-phosphate. The glyceraldehyde 3-phosphate is converted to pyruvate in the bottom portion of the glycolytic pathway.

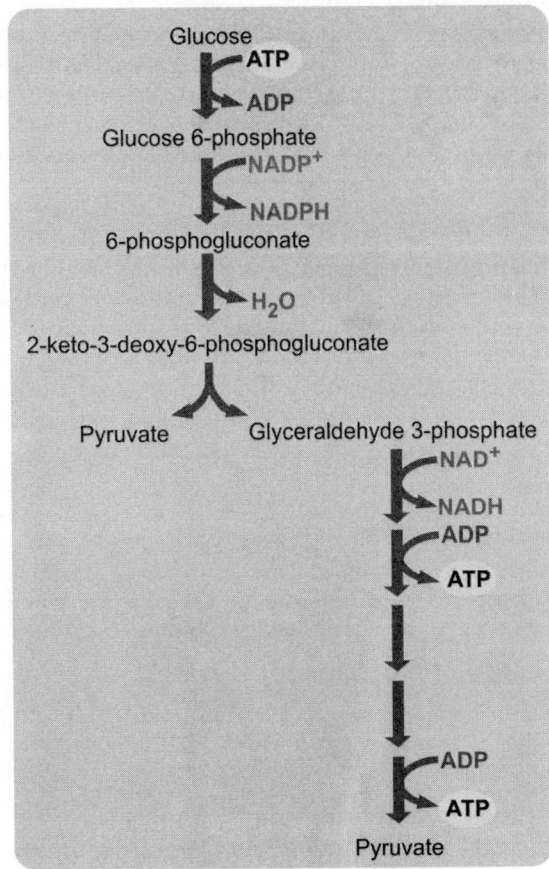

*Fig. 109 : **ED pathway.***

If the Entner-Doudoroff pathway degrades glucose to pyruvate in this way, it yields one ATP, one NADPH, and one NADH per glucose metabolized.

Most bacteria have the glycolytic and pentose phosphate pathways, but some substitute the Entner-Doudoroff pathway for glycolysis. The Entner-Doudoroff pathway is generally found in *Pseudomonas, Rhizobium, Azotobacter, Agrobacterium*, and a few other gram-negative genera. Very few gram-positive bacteria have this pathway, with *Enterococcus faecalis* being a rare exception.

GLUCONEOGENESIS

The synthesis of glucose from non-carbohydrate precursors is known as gluconeogenesis. The major site of gluconeogenesis is liver. It usually occurs when the carbohydrate in the diet is insufficient to meet the demand in the body, with the intake of protein rich diet and at the time of starvation, when tissue proteins are broken down to amino acids.

Gluconeogenesis and glycolysis

Gluconeogenesis and glycolysis are opposing metabolic pathways and share a number of enzymes. In glycolysis, glucose is converted to pyruvate and in gluconeogenesis pyruvate is converted to glucose. However gluconeogenesis is not exact reversal of glycolysis. There are three essentially irrevesible steps in glycolysis which are In gluconeogenesis these three reactions are bypassed or substituted by the following news ones.

Reactions of gluconeogenesis

1. The formation of phosphoenol pyruvate begins with the carboxylation of pyruvate at the expense of ATP to form oxalo acetate.Oxaloacetate is converted to phosphoenolpyruvate by phosphorylation with GTP, accompanied by a simultaneous decarboxylation.

2. Fructose 6-phosphate is formed from fructose 1,6-diphosphate by hydrolysis and the enzyme fructose 1,6-diphosphatase catalyses this reaction.

3. Glucose is formed by hydrolysis of glucose 6-phosphate catalysed by glucose 6-phosphatase.

Gluconeogenesis of amino acids

Amino acids which could be converted to glucose are called glucogenic amino acids. Most of the glucogenic amino acids are converted to the intermediates of citric acid cycle either by transamination or deamination.

Gluconeogenesis of Propionate

Propionate is a major source of glucose in ruminants, and enters the main gluconeogenic pathway via the citric acid cycle after conversion to succinyl CoA.

Gluconeogenesis of Glycerol

At the time of starvation glycerol can also undergo gluconeogenesis. When the triglycerides are hydrolysed in the adipose tissue, glycerol is released.

Further metabolism of glycerol does not take place in the adipose tissue because of the lack of glycerol kinase necessary to phosphorylate it. Instead, glycerol passes to the liver where it is phosphorylated to glycerol 3-phosphate by the enzyme glycerol kinase.

This pathway connects the triose phosphate stage of glycolysis, because glycerol 3-phosphate is oxidized to dihydroxy acetone phosphate in the presence of NAD^+ and glycerol 3-phosphate dehydrogenase.

This dihydroxy acetone phosphate enters gluconeogenesis pathway and gets converted to glucose. Liver and kidney are able to convert glycerol to blood glucose by making use of the above enzymes.

THE GENERATION OF ATP

ATP is generated through a process called phosphorylation. It is a chemical process in which a phosphate group is added to an organic molecule (ADP). In living cells phosphorylation is associated with respiration. The energy released during metabolic or photosynthetic processes is captured in the energy-rich phosphate bonds of certain molecules, most commonly in the high-energy bonds of adenosine triphosphate (ATP). Micro organisms use three mechanisms of phosphorylation to generate ATP from ADP. They are *Substrate-Level Phosphorylation, Oxidative Phosphorylation and Photo Phosphorylation.*

Substrate-Level Phosphorylation

In substrate-level phosphorylation, ATP is usually generated when a high-energy is directly transferred from a phosphorylated compound (a substrate) to ADP.

Oxidative Phosphorylation

In **oxidative phosphorylation,** electrons are transferred from organic compounds to one group of electron carriers (usually to NAD⁻ and FAD). Then, the electrons are passed through a series of different electron carriers to molecules of oxygen (O_2) or other oxidized inorganic and organic molecules. This process occurs in the plasma membrane of prokaryotes and in the inner mitochondrial membrane of eukaryotes. The sequence of electron carriers used in oxidative phosphorylation is called an **electron transport chain (system).** The transfer of electrons from one electron carrier to the next releases energy, some of which is used to generate ATP from ADP through a process called *chemiosmosis.*

Photophosphorylation

This occurs only in photosynthetic cells, which contain light-trapping pigments such as chlorophylls. In photosynthesis, organic molecules, especially sugars, are synthesized with the energy of light from the energy-poor building blocks carbon dioxide and water. Photophosphorylation starts this process by converting light energy to the chemical energy of ATP and NADPH, which, in turn, are used to synthesize organic molecules. As in oxidative phosphorylation, an electron transport chain is involved.

Electron transport chain

An **electron transport chain (system)** consists of a sequence of carrier molecules that are capable of oxidation and reduction. As electrons are passed through the chain, there occurs a stepwise release of energy, which is used to drive the chemiosmotic generation of ATP. In eukaryotic cells, mitochondria is the site for electron transport chain; in prokaryotic cells, it is found in the plasma membrane.

There are three classes of carrier molecules in electron transport chains. The first are **flavoproteins.** These proteins contain flavin, a coenzyme derived from riboflavin (vitamin B_2), and are capable of performing alternating oxidations and reductions. One important flavin coenzyme is flavin mononucleotide (FMN). The second class of carrier molecules are **cytochromes,** proteins with an iron-containing group (heme) capable of existing alternately as a reduced form (Fe^{2+}) and an oxidized form (Fe^{3+}). The cytochromes involved in electron transport chains include cytochrome *b* (cyt *b*), cytochrome *c1* (cyt *c1*), cytochrome *c* (cyt *c*), cytochrome *a* (cyt *a*), and cytochrome *a3* (cyt *a3*). The third class is known as **ubiquinones,** or **coenzyme Q,** these are small nonprotein carriers.

The first step in the mitochondrial electron transport chain involves the transfer of high-energy electrons from NADH to FMN.

ELECTRON TRANSPORT CHAIN

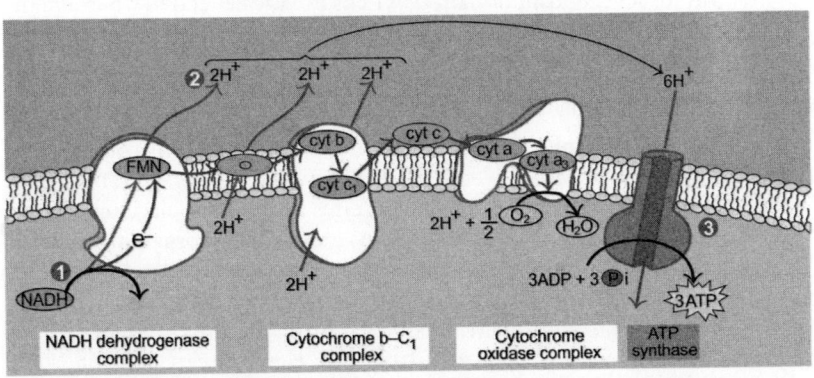

*Fig. 110 : **Electron transport chain.***

This transfer actually involves the passage of a hydrogen atom with two electrons to FMN, which then picks up an additional H^+ from the surrounding aqueous medium. As a result of the first transfer, NADH is oxidized to NAD^+, and FMN is reduced to $FMNH_2$. In the second step in the electron transport chain, $FMNH_2$ passes $2H^+$ to the other side of the mitochondrial membrane and passes two electrons to Q. As a result, $FMNH_2$ is oxidized to FMN. Q also picks up an additional $2H^+$ from the surrounding aqueous medium and releases it on the other side of the membrane.

The next part of the electron transport chain involves the cytochromes. Electrons are passed successively from Q to cyt *b*, cyt *c1*, cyt *c*, cyt *a*, and cyt *a3*. Each cytochrome in the chain is reduced as it picks up electrons and is oxidized as it gives up electrons. The last cytochrome, cyt a_3, passes its electrons to molecular oxygen (O_2), which becomes negatively charged and then picks up protons from the surrounding medium to form H_2O.

$FADH_2$ is derived from the Krebs cycle. $FADH_2$ adds its electrons to the electron transport chain at a lower level than NADH. Because of this, $FADH_2$ produces 2 ATP whereas NADH generats three ATP.

Electron flow down the chain is accompanied at several points by the active transport (pumping) of protons from the matrix side of the inner mitochondrial membrane to the opposite side of the membrane. The result is a buildup of protons on one side of the membrane. Just as water behind a dam stores energy that can be used to generate electricity, this buildup of protons provides energy for the generation of ATP by the chemiosmotic mechanism.

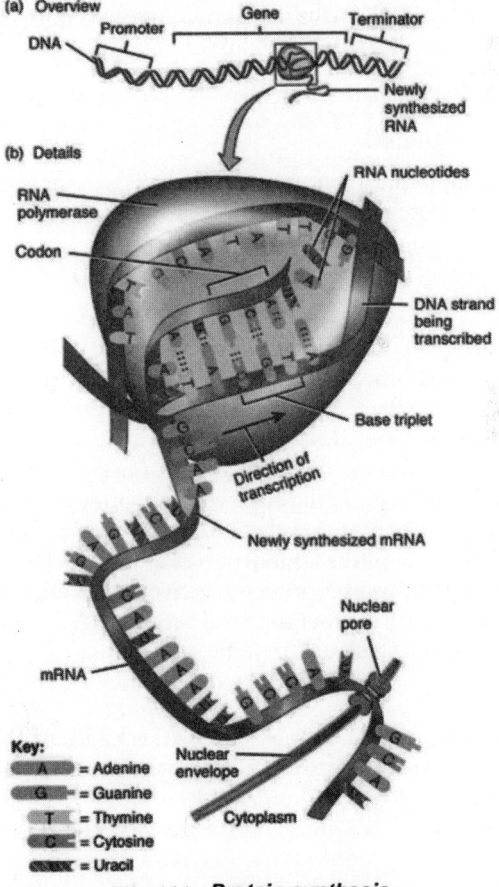

Fig. 111 : Protein synthesis.

The Chemiosmotic Mechanism of ATP Generation

The mechanism of ATP synthesis using the electron transport chain is called **chemiosmosis.**

As energetic electrons from NADH (or chlorophyll) pass down the electron transport chain, some of the carriers in the chain pump—actively transport—protons across the membrane. Such carrier molecules are called *proton pumps.* The phospholipid membrane is normally impermeable to protons, so this one-directional pumping establishes a proton gradient (a difference in the concentrations of protons on the two sides of the membrane). In addition to a concentration gradient, there is an electrical charge gradient. The excess H+ on one side of the membrane makes that side positively charged compared with the other side. The resulting electrochemical gradient has potential energy, called the *proton motive force.*

The protons on the side of the membrane with the higher proton concentration can diffuse across the membrane only through special protein channels that contain an enzyme called *ATP synthase.* When this flow occurs, energy is released and is used by the enzyme to synthesize ATP from ADP

PROTEIN METABOLISM

Protein biosynthesis

In a protein molecule amino acids are joined together by peptide bonds. In the process of protein synthesis also known as translation of mRNA, the amino acids are added sequentially in a specific number. The protein synthesizing mechanism involves the following steps:

Transcription

The formation of RNA complementary to a DNA strand is called transcription. In this process, the RNAs required for protein synthesis are synthesized on DNA strands. This reaction is catalyzed by the enzyme RNA polymerase. The enzyme, RNA polymerase I, II, III are involved in the synthesis of rRNA (ribosomal RNA) mRNA (messenger RNA) and tRNA (transfer RNA) respectively in the eukaryotes. In prokaryotes only one type of RNA polymerase is present to synthesize all the three classes of RNA. In the DNA double helix, one of the strands serves as a template to produce RNA. The RNA produced by transcription is inactive and is called pre-RNA. They become active after further processing. All these RNA are processed through chemical reactions and structural modifications.

Translation

Translation is a process by which the base sequence of DNA transcribed to the mRNA is interpreted into amino acid sequence of a polypeptide chain.

Translation involves the following steps:

1. Activation of amino acid
2. Transfer of activated amino acid to tRNA

3. Initiation of polypeptide chain

4. Elongation of polypeptide chain

5. Termination of polypeptide chain

Activation of amino acid

Amino acids, the building blocks of proteins. They are present in the cytoplasm. They are activated before they are transported by tRNA. The amino acids are activated by ATP with the help of the enzyme amino acyl synthetase. Amino acyl synthetase is specific in activating each amino acid. The activated amino acid is called amino acyl adenylate or amino acyl AMP. Pyrophosphate is released.

$$AA + ATP \longrightarrow aminoacyl\ AMP$$

Transfer of activated amino acid to tRNA

The same enzyme that activates the amino acid catalyses its transfer to a molecule of transfer RNA at the 3′ hydroxyl of the ribose, an ester with a high potential for group transfer. In this reaction AMP and the enzyme amino acyl synthetase are released.

Initiation of polypeptide chain

Protein synthesis is initiated by the selection and transfer of the first amino acid into ribosomes. This process requires ribosome subunits, amino acyl tRNA complex, mRNA and initiation factors (IF). Initiation of polypeptide chain involves the following steps.

1. The 30s ribosomal subunit attaches to the 5′ end of the mRNA to form an mRNA 30s complex. This process requires the initiation factor IF-3 and Mg^{2+} ions. The attachment is made at the first codon of the mRNA.

2. The first codon of mRNA will be always AUG. This codon specifies the amino acid methionine. So the first amino acid in the synthesis of any polypeptide chain is methionine.

3. The tRNA having the anticodon UAC (complementary to AUG) transports methionine to the 30s ribosome and attaches itself to the initiation codon on mRNA. The tRNA, mRNA and 30s ribosome subunit form a complex called 30s - pre initiation complex. This process requires initiation factors and GTP.

4. 30s-pre initiation complex joins with 50s ribosomal subunit to form initiation complex. The initiation complex is formed of 70s ribosome, mRNA and met -RNA (methionine RNA).

5. The 70s ribosome has two slots for the entry of amino acyl tRNA, namely P site (peptidyl site) and A site (amino acid site). The first tRNA i.e. met RNA is attached to the P site of 70s ribosome.

Elongation of polypeptide chain

Elongation refers to sequential addition of amino acids to methionine, as per the sequence of codon in the mRNA. It involves the following steps:

1. The second codon in the mRNA is recognised and as per the recognition, the amino acyl tRNA containing the corresponding anticodon moves to the 70s ribosome and fits into the A-site. Here the anticodon of tRNA base pairs with the second codon of mRNA.

2. A peptide bond is formed between the carboxyl group (–COOH) of first amino acid of site P and the amino group (–NH$_2$) of second amino acid of A-site. The peptide bond links two amino acids to form a dipeptide. The bonding is catalysed by the enzyme peptidyl transferase which is present in 50s ribosomal subunit.

3. After the formation of peptide bond, the methionine and tRNA are separated by an enzyme called tRNA deacylase.

4. The dissociated tRNA is then released from P-site into the cytoplasm for further amino acylation.

5. Now the ribosome moves on the mRNA in the 5′→3′ direction so that the first codon goes out of ribosome, the second codon comes to lie in the P-site from A-site and the third codon comes to lie in the A-site. Simultaneously, the second tRNA is shifted from A-site to P-site. All these events, the movement of ribosome, the release of first tRNA from P-site and shifting of second tRNA from A-site to P-site constitute *translocation*. Translocation is catalyzed by the enzyme *translocase*.

6. The third codon is recognised and the amino acyl tRNA containing the corresponding anticodon moves to the 70s ribosome and fits into the A-site. The anticodon base pairs with the codon. A peptide bond is formed between the third amino acid of site-A and the second amino acid of the dipeptide present in the P-site. Thus a tripeptide is formed.

7. The amino acids are added one by one as per the codon in the mRNA and hence the tripeptide is converted into polypeptide chain. The polypeptide chain elongates by the addition of more and more amino acids.

8. The elongation of polypeptide chain is brought about by a number of protein factors called elongation factors.

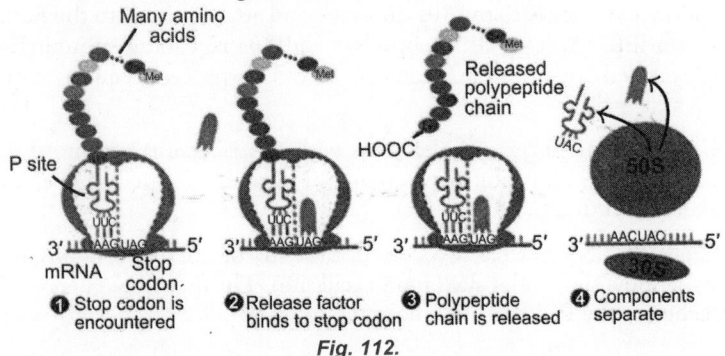

Fig. 112.

Termination of polypeptide chain

Termination is the completion of polypeptide chain. By termination, a polypeptide chain is finished and released. The polypeptide chain is **completed, when the ribosome reaches the 3′ end of mRNA.** The 3′ end contains stop codons or termination codons. They are UAG or UAA or UGA. Termination is helped by the terminating protein factors. The terminated polypeptide chain is released from the ribosome.

After the release of polypeptide chain, the 70s unit dissociates into 50s and 30s sub-units. These subunits are again used in the formation of another initiation complex.

The polypeptide chain released after translation is inactive. It is processed to make it active. In the processing the initiating amino acid methionine is removed. Along with methionine a few more amino acids are removed from the N-terminal of the polypeptide. The processing is carried out by deformylase and amino peptidase. This processing is called as post translational modifications.

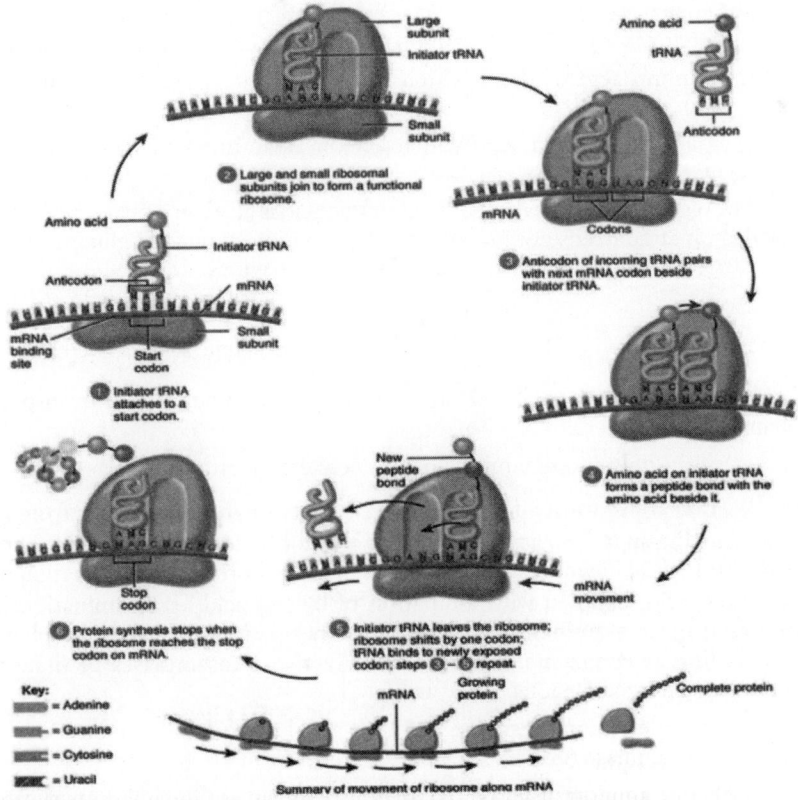

Fig. 113.

CATABOLISM OF AMINO ACIDS

Amino acid follows its own specific metabolic pathway, a few general reactions are found to be common in the catabolism of nearly all the amino acids. Most of the amino acids are converted to a-keto acids by the removal of nitrogen in the form of ammonia which is quickly transformed into urea or it gets incorporated into some other amino acids.

1. Oxidat ive deaminat ion : Deamination means removal of the amino groups from amino acids. This is the mechanism here in the amino acids lose two hydrogen atoms (dehydrogenation) to form keto acids and ammonia. Oxidative deamination is accompanied by oxidation and is catalysed by specific amino acid oxidases or more appropriately, dehydrogenases present in liver and kidneys. The process of oxidative deamination takes place in two steps. The first step is oxidation (dehydrogenation) of amino acid resulting in the formation of imino acid. The imino acid then undergoes the second step, namely hydrolysis which results in a keto acid and ammonia.

$$\text{Amino acid} \rightarrow \text{Imino acid} \rightarrow \text{Keto acid} \rightarrow \text{Ammonia}$$

The first reaction is catalyzed by amino acid oxidase (also called dehydrogenase) and the coenzyme FAD or FMN takes up the hydrogen. There are two types of amino acid oxidases depending upon the substrate on which they act, namely,

1. L-amino acid oxidases which act on L-amino acids (FMN acts as coenzyme).

2. D-amino acid oxidases which act on D-amino acids (FAD acts as coenzyme).

The oxidative deamination of L-glutamic acid is an exceptional case where the deamination needs not only the zinc-containing enzyme L-glutamic acid dehydrogenase but also NAD^+ or $NADP^+$ as coenzymes.

$$\text{L-Glutamic acid} + NAD^+ \xrightarrow{\substack{\text{Glutamic acid} \\ \text{dehydrogenase}}} \alpha\text{-Iminoglutamic acid} + NADH + H^+$$

NADH gets oxidized to NAD^+ as it passes through the electron transport chain.

$$\alpha\text{-Iminoglutamic acid} \xrightarrow{\text{HO}} \alpha\text{-keto glutaric acid.}$$

2. Transaminat ion : The process of transfer of an amino group from an amino acid to an α-keto acid, resulting in the formation of a new amino acid and keto acid is known as transamination. In other words, it is deamination of an amino acid, coupled with amination of α-keto acid. Transamination is catalyzed by transaminases or aminotransferases with pyridoxal phosphate functioning as coenzyme. There are two active transaminases in tissues, catalyzing interconversions. They are

1. Aspartate aminotransferase (AST) is also known as Glutamate - oxalo acetate transaminase (GOT)

2. Alanine aminotransferase (ALT) is also known as Glutamate- pyruvate transaminase (GPT).

It catalyses the transfer of NH_2 group from glutamate to pyruvate, resulting in the formation of α-ketoglutaric acid and alanine.

$$\begin{array}{ccc}
\text{L-Glutamate} & & \text{α-keto glutarate} \\
+ & \xrightarrow{\quad AST \quad} & + \\
\text{oxalo acetate} & & \text{L-asparatate}
\end{array}$$

$$\begin{array}{ccc}
\text{L-Glutamate} & & \text{α-keto glutarate} \\
+ & \xrightarrow{\quad ALT \quad} & + \\
\text{Pyruvate} & & \text{Alanine}
\end{array}$$

3. Decarboxylat ion : This refers to the removal of CO_2 from the carboxyl group of amino acids. The removal of CO_2 needs the catalytic action of enzymes decarboxylases and the pyridoxal phosphate coenzyme. The enzymes act on amino acids resulting in the formation of the corresponding amines with the liberation of CO_2.

$$\text{Amino acid} \xrightarrow{\overset{\text{Amino acid}}{\text{decarboxylate}}} \text{Amine} + CO_2$$

4. Transmet hylation : The transfer of methyl group from one compound to another is called transmethylation and the enzymes involved in the transfer are known as transmethylases.

Biosynthesis of Aminoacids

Amino acid synthesis is the set of biochemical processes by which the various amino acids are produced from other compounds. The substrates for these processes are various compounds in the organism's diet or growth media. Not all organisms are able to synthesise all amino acids. All amino acids are derived from intermediates in glycolysis, the citric acid cycle, or the pentose phosphate pathway. Nitrogen enters these pathways by way of glutamate and glutamine. Different organisms vary greatly in their ability to synthesize the 20 amino acids. A useful way to organize the amino acid biosynthetic pathways is to group them into 6 families corresponding to the metabolic precursor of each amino acid.

α-Ketoglularate	Oxaloacetate	Phosphoenolpyruvate and erythrose-4-phosphate
Glutamate	Aspartate	Tryptophan*
Glutamine	Asparagine	Phenylalanine*
Proline	Methionine*	Tyrosine
Argine	Threonine*	
	Lysine	
	Isoleucine	

3-Phosphoglycerate	Pyruvate	Ribose-5-phosphate
Serine	Alanine	Histidine*
Glyceine	Valine*	
Cysteine	Leucine	

α-Ketoglutarates

The α-ketoglutarate family of amino acid synthesis (synthesis of glutamate, glutamine, proline and arginine) begins with α-ketoglutarate, an intermediate in the Citric Acid Cycle.

Erythrose 4-phosphate and phosphoenolpyruvate

Phenylalanine, tyrosine, and tryptophan are known as the aromatic amino acids. The synthesis of all three share a common beginning to their pathways; the formation of chorismate from phosphoenolpyruvate (PEP) and erythrose 4- phosphate (E4P).

Oxaloacetate/aspart ate

The oxaloacetate/aspartate family of amino acids is composed of lysine, asparagine, methionine, threonine and isoleucine. Aspartate can be converted into lysine, asparagine, methionine and threonine. Threonine also gives rise to isoleucine.

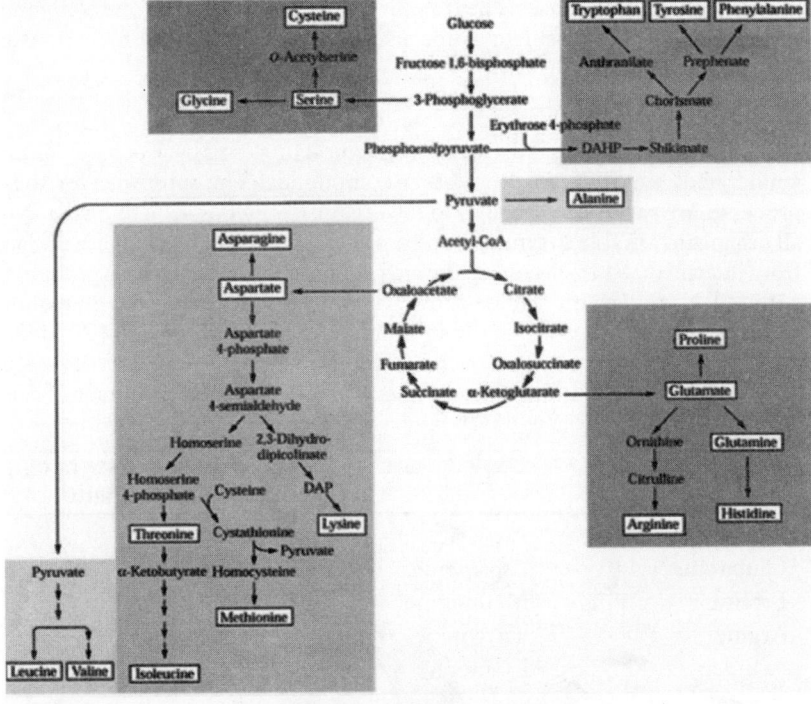

*Fig. 114 : **Aminoacid biosynthesis.***

Aspartate: The enzyme aspartokinase, which catalyzes the phosphorylation of aspartate and initiates its conversion into other amino acids, can be broken up into 3 isozymes, AK-I, II and III. AK-I is feed-back inhibited by threonine, while AK-II and III are inhibited by lysine. As a sidenote, AK-III catalyzes the phosphorylation of aspartic acid that is the commitment step in this biosynthetic pathway. The higher the concentration of threonine or lysine, the more aspartate kinase becomes downregulated. (AK-Aspartate kinase)

Lysine: Lysine is synthesized from aspartate via the diaminopimelate (DAP) pathway. The initial two stages of the DAP pathway are catalyzed by aspartokinase and aspartate semialdehyde dehydrogenase and play a key role in the biosynthesis of lysine, threonine and methionine.

Asparagine: There are two different asparagine synthetases found in bacterial species. These two synthetases, which are both referred to as the AsnC protein, are coded for by two genes: AsnA and AsnB.

Methionine: Methionine synthesis is under tight regulation. The repressor protein MetJ, in cooperation with the corepressor protein S-adenosyl-methionine, mediates the repression of methionine's biosynthetic pathway.

Threonine: The biosynthesis of threonine is regulated via allosteric regulation of its precursor, homoserine, by structurally altering the enzyme homoserine dehydrogenase. This reaction occurs at a key branch point in the pathway, with the substrate homoserine serving as the precursor for the biosynthesis of lysine, methionine, threonine and isoleucine. High levels of threonine result in low levels of homoserine synthesis. The synthesis of aspartate kinase (AK), which catalyzes the phosphorylation of aspartate and initiates its conversion into other amino acids, is feed-back inhibited by lysine, isoleucine, and threonine, which prevents the synthesis of the amino acids derived from aspartate. Isoleucine: The enzymes threonine deaminase, dihydroxy acid dehydrase and transaminase are controlled by end-product regulation.

Ribose 5-phosphates

The synthesis of histidine in *E. coli* is a complex pathway involving 10 reactions and 10 enzymes. Synthesis begins with 5-phosphoribosyl-pyrophosphate (PRPP) and finishes with histidine and occurs through the reactions of the following enzymes.

HisG → HisE/HisI → HisA → HisH → HisF → HisB → HisC → HisB → HisD (HisE/I and HisB are both bifunctional enzymes).

3-Phosphoglycerates

Serine: Serine is the first amino acid in this family to be produced; it is then modified to produce both glycine and cysteine (and many other biologically important molecules). Serine is formed from 3-phosphoglycerate in the following pathway:

3-phosphoglycerate → phosphohydroxyl-pyruvate → phosphoserine → serine

The conversion from 3-phosphoglycerate to phosphohydroxyl-pyruvate is achieved by the enzyme phosphoglycerate dehydrogenase.

Glycine : Glycine is synthesized from serine using the enzyme serine hydromethyltransferase (SHMT), which is coded by the gene glyA. The enzyme effectively removes a hydroxyl group from serine and replaces it with a methyl group to yield glycine. This reaction is the only way *E. coli* can produce glycine.

Cysteine : Cysteine is a very important molecule for a bacterium's survival. This amino acid harbors a sulfur atom and can actively participate in disulfide bond formation.

Pyruvates

Pyruvate is the end result of glycolysis and can feed into both the TCA cycle and fermentation processes. Reactions beginning with either one or two molecules of pyruvate cause the synthesis of alanine, valine, and leucine.

Alanine : Alanine is produced by the transamination of one molecule of pyruvate using two alternate steps: (1) conversion of glutamate to α-ketoglutarate using a glutamate-alanine transaminase, and (2) conversion of valine to α-ketoisovalerate via Transaminase C.

Valine : Valine is produced by a four-enzyme pathway. It begins with the reaction of two pyruvate molecules catalyzed by Acetohydroxy acid synthase yielding α-acetolactate. Step two is the $NADPH^+ + H^+$ - dependent reduction of α-acetolactate and migration of the methane groups to produce, α,β-dihydroxyisovalerate. This is catalyzed by Acetohydroxy isomeroreductase. The third reaction is the dehydration reaction of, α,β-dihydroxyisovalerate catalyzed by Dihydroxy acid dehydrase resulting in β-ketoisovalerate. Finally, a transamination catalyzed either by an alanine-valine transaminase or a glutamate-valine transaminase results in valine.

Leucine : The leucine synthesis pathway diverges from the valine pathway beginning with α-ketoisovalerate. α-Isopropylmalate synthase reacts with this substrate and Acetyl CoA to produce α-isopropylmalate. An isomerase then isomerizes α-isopropylmalate to β-isopropylmalate. The third step is the NAD+ dependent oxidation of â-isopropylmalate via the action of a dehydrogenase to yield β-ketoisocaproate. Finally is the transamination via the action of a glutamate-leucine transaminase to result in leucine

UREA CYCLE

Living organisms excrete the excess nitrogen resulting from the metabolic breakdown of amino acids in one of three ways. Many aquatic animals simply excrete ammonia. Where water is less, plentiful processes have evolved that convert ammonia to less toxic waste products which require less water for excretion. One such product is urea, which is excreted by most terrestrial vertebrates; another is uric acid, which is excreted by birds and terrestrial reptiles.

Accordingly, living organisms are classified as being either ammonotelic (ammonia excreting), urotelic (urea excreting) and uricotelic (uric acid excreting). Some animals can shift from ammonotelism to urotelism or uricotelism if their water supply becomes restricted. Urea is synthesised in the liver by the enzymes of the urea cycle. It is then secreted into the blood stream and sequestered by the kidneys for excretion in the urine.

The urea cycle reactions were elucidated by Hans Krebs and Kurt Henseleit. This cycle starts with the amino acid ornithine. The cycle is confined only to the mitochondria and cytoplasm of the cells of liver and it is found that the enzyme, arginase which is required in the final step of urea formation is present only in the liver and absent in all the other tissues.

Urea cycle occurs partially in the mitochondria and partially in the cytosol with ornithine and citrulline being transported across the mitochondrial membrane by specific membrane systems. The following are the various reactions in the process of urea formation.

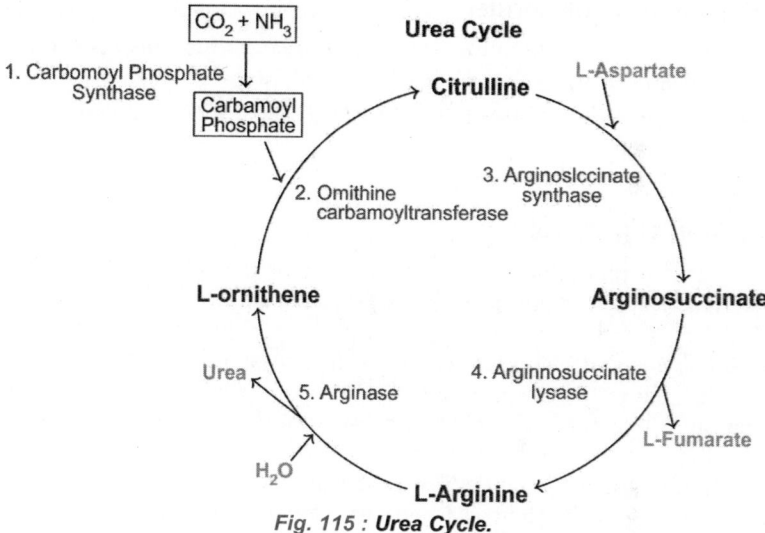

Fig. 115 : *Urea Cycle.*

1. **Carbamoyl phosphate formation** : Carbamoyl phosphate synthetase catalyses the condensation and activation of NH_4^+ and HCO_3^- to form carbamoyl phosphate.

2. **Citrulline formation from ornithine** : Ornithine transcarbamylase transfers the carbamoyl group of carbamoyl phosphate to ornithine, yielding citrulline. The reaction occurs in the mitochondria so that ornithine, which is produced in the cytosol, must enter the mitochondria via a specific transport system. Like wise, since the remaining urea cycle reactions occur in the cytosol, citrulline must be transported from the mitochondria.

3. **Argininosuccinate formation** : Citrulline undergoes condensation with amino group of aspartate to form arigininosuccinate this reaction requires ATP, Mg^{2+} and the enzyme argininosuccinate synthetase.

4. **Formation of arginine and fumarate** : The enzyme argininosucccinase catalyses the elimination of arginine from the aspartate carbon skeleton forming fumarate.

5. **Formation of urea :** The fifth and the final reaction in the urea cycle is the hydrolysis of arginine by the enzyme arginase to yield urea and ornithine. Ornithine is then returned to the mitochondria for another round of th e cycle.

LIPID METABOLISM

Lipids plays an important role in metabolism as a fuel for the production of ATP. The first step in lipid metabolism is hydrolysis of lipid in to fatty acids and glycerol through the action of lipolytic enzymes.

$$\text{Fat} \xrightarrow[3H_2O]{\text{Lipase}} \text{Glycerol} + 3\,\text{Fatty acids}$$

Oxidat ion of Triglycerides

Triglycerides are generally used by chemoorganotrophic microorganisms. Lipase enzyme hydrolyses triglyceride in three steps of reactions with the uptake of 3 molecule of water. This hydrolysis releases glycerol and fatty acids.

$$\text{Fat} \xrightarrow[3H_2O]{\text{Lipase}} \text{Glycerol} + 3\,\text{Fatty acids}$$

Oxidation of glycerol

Glycerol is phosphorylated by making use of enzyme glycerokinase and converted to glycerol 3 phosphate in the presence of ATP. Glycerol 3 phosphate is dehydrogenated and form dihydroxy acetone phosphate in the presence of NAD. Dihydroxy acetane phosphate undergone isomerization process with the aid of Triose phosphate isomerases and converted to glyceroldehyde 3 phosphate. Glyceroldehyde 3 phosphate undergoes forward or reverse reactions of glycolysis. Forward reactions leads to the formation of pyruvic acid. Reverse reactions leads to the formation of glucose. Condensation of dihydroxy acetone phosphate and glyceroldehyde 3 phosphate form fructose 1, 6 di phosphate, which enters reverse glycolysis and form glucose.

Oxidation of fatty acids

Fatty Acids are the major source of energy. They are oxidized to CO2 and water with the liberation of energy. Fattyacids oxidation takes place in mitochondria in eukaryotes and cytoplasm in prokaryotes. Three major theories explain the process of fatty acid oxidation. They are α-oxidation, β-oxidation and ω-oxidation.

α oxidation

It is a process of fattyacids breakdown in the carboxyl end with the removal of a single carbon from α position. This theory was proposed by P. K. Stump. Oxidation is takes place in peroxisomes to break down dietary phytanic acid. Phytanic acid cannot undergoes oxidation due to its α methyl branch. Phytanic acid is converted to pristanic acid then acquire acetyl CoA and subsequently become oxidized. Phytanic acid is attached to CoA to form phytanoyl CoA.

Phytanoyl CoA is oxidized by phytanoyl CoA dioxygenase and form hydroxy phytanoyl CoA. Hydroxy phytanoyl CoA is cleaved by lyase enzyme and form pristanal and formyl CoA. Pristanol is oxidized by aldehyde dehydrogenase to form pristanic acid. Pristanic acid may involve in β oxidation.

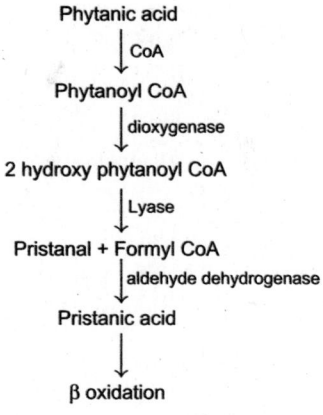

Fig. 116 : α-oxidation.

β oxidation

This theory was proposed by knoop in 1904. This theory stated that oxidation of fatty acid occurs at the β carbon results in the formation of acetone from a terminal two carbon and leaving behind a shorter fatty acid containing two carbon atom lesser than the original. This oxidation provides larger amount of energy. This oxidation takes place in five steps. They are:

1. Activation, 2. Desaturation, 3. Hydration, 4. Oxidation and 5. Thiolysis.

1. Act ivat ion : Fattyacids are activated by the addition of CoA, ATP and Magnesium ions with the help of thiokinase. Activated fatty acid is known as fatty acyl CoA.

Fatty acid + CoA + ATP ⟶ Acetyl CoA derivative

2. Desaturation : Activated fatty acid dehydrogenated by acetyl; CoA dehydrogenase in to α,β factty acid CoA derivative.

Acetyl CoA ⟶ α,β fatty acid CoA

3. Hydration : α,β, fatty acid CoA undergoes hydration and combined with a molecule of water using an enzyme hydratase and form α hydroxyl acyl CoA

α,β fatty acid CoA + Water ⟶ α hydroxyl acyl CoA

4. Oxidat ion : β hydroxyl acyl CoA derivative undergoes oxidation to form keto fatty acid in the presence of dehydragenase and NAD.

β hydroxyl acyl CoA + NAD ⟶ α,β keto fatty acid CoA

5. Thiolysis : α keto fatty acid CoA is split into acetyl CoA and active fatty acid with the help of thiolase.

α keto fatty acid CoA ⟶ Acetyl CoA + active fatty acid

The newly formed acetyl CoA may enter in TCA cycle. Active fatty acid acyl CoA reenters the cycle and repeated till the fatty acid is completely splitup into acetyl units.

α keto fatty acid CoA ————→ α Acetyl CoA + active fatty acid

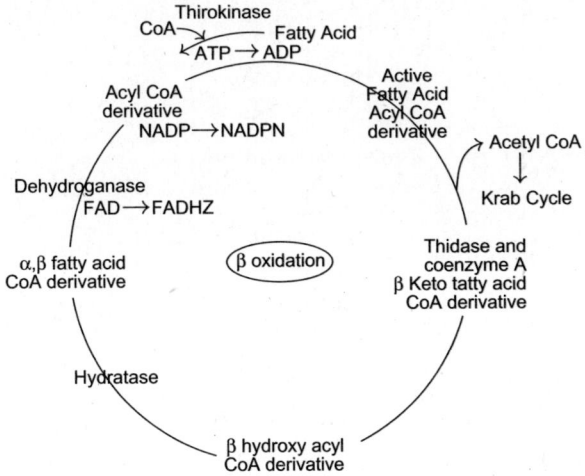

Fig. 117 : *β Oxidation.*

ω oxidation

This theory was proposed by Verkade. This theory stated that oxidation occurs at the carbon atom situated farthest from the carboxyl group [ω carbon]. Omega oxidation results in the formation of Dicarboxylic acid. After this process, α oxidation starts at both ends successively. ω oxidation is quicker than regular ω oxidation process. It occurs in three steps.

1. First step introduces hydroxyl group in to the ω carbon. It is done by mixed function oxidase.

2. Oxidation of the hydroxyl group to an aldehyde by NAD. It is by alcohol dehydrogenase.

3. Oxidation of the aldehyde group by aldehyde dehydrogenase to a carboxylic acid by NAD.

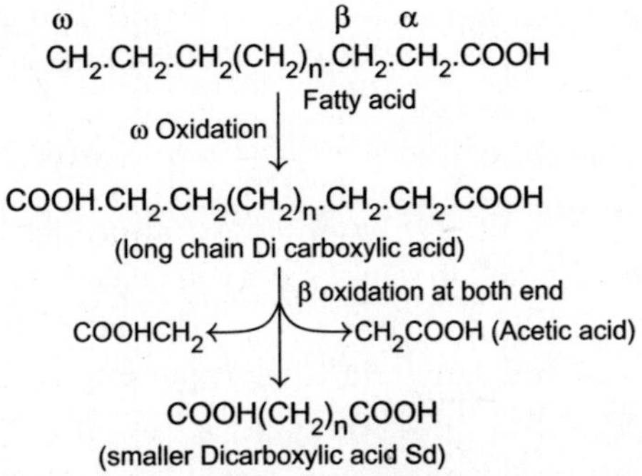

$$\overset{\omega}{CH_2}.CH_2.CH_2(CH_2)_n.\overset{\beta}{CH_2}.\overset{\alpha}{CH_2}.COOH$$

Fatty acid

ω Oxidation ↓

$$COOH.CH_2.CH_2(CH_2)_n.CH_2.CH_2.COOH$$

(long chain Di carboxylic acid)

β oxidation at both end

$$COOHCH_2 \longleftarrow \diagdown\diagup \longrightarrow CH_2COOH \text{ (Acetic acid)}$$

↓

$$COOH(CH_2)_nCOOH$$

(smaller Dicarboxylic acid Sd)

Fig. 118 : *ω oxidation.*

Synthesis of fatty acids

The process of synthesizing fatty acids are called lipogenesis. It takes place in the cytoplasm. Fatty acid synthesis occurs in all organisms. Multienzyme complex called fattyacid synthetase is responsible for fatty acid synthesis.

- Melonyl CoA reacts with acetyl CoA and gets decarboxylated in the presence of condensing enzyme to form aceto acetyl CoA.
- Aceto acetyl CoA is reduced to â hydroxy butyryl CoA in the presence of an enzyme ketoacyl CoA reductase.
- β hydroxy butyryl CoA, by the removal of water molecule, converted to α,β unsaturated butyryl CoA in the presence of hydratase.
- α,β unsaturated butyryl CoA is reduced to butyryl CoA by the action of acyl CoA reductase.
- Final step of first cycle is the formation of butyric acid (4 Carbon) by the action of enoyl ACP reductase.

In the second step butyryl ACP condenses with melonyl ACP with the similar sequence of reactions results in 6 Carbon ACP. Similar cycle follows until 16 carbon ACP is formed. 16 carbon fatty acid is palmitic acid.

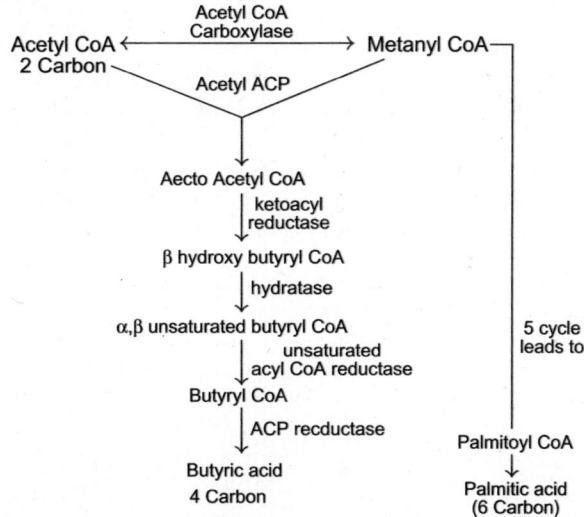

*Fig. 119 : **Fatty acid Synthesis.***

Biosynthesis of triglycerides

Fatty acids combine with glycerol to form triglycerides. The glycerol can be derived from dihydroxy acetone phosphate. It is an energy yielding process. The first step in triglyceride synthesis is the activation of glycerol and fatty acids by ATP. Glycerol is activated by ATP in the presence of glycerokinase to form   glycerol phosphate. Fatty acid is activated to form acyl CoA by the enzyme thiokinase in the presence of ATP, CoA and SH. The glycerol phosphate is esterified with 2 molecules of aceyl CoA to form α,β diglyceride phosphate (phosphatidic acid) in the presence of glycerol phosphate acyl transferase. The phosphatidic acid is dephosphorylated by the presence of phosphatase to form α,β diglyceride. Another one molecule of acyl CoA is esterified with the diglyceride, to form triglyceride.

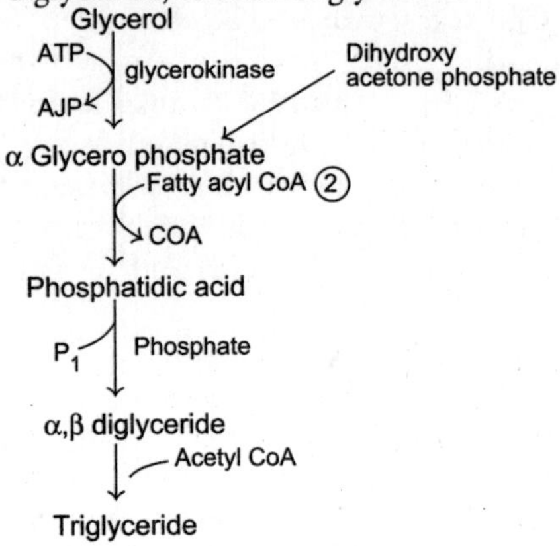

Glycerol

ATP — glycerokinase — Dihydroxy acetone phosphate

AJP

α Glycero phosphate

Fatty acyl CoA ②

COA

Phosphatidic acid

P_1 — Phosphate

α,β diglyceride

Acetyl CoA

Triglyceride

Fig. 120 : Triglyceride synthesis.

Biosynthesis of cholesterol

Cholesterol is synthesized from 2 carbon units in the form of acetyl CoA. Two molecules of acetyl CoA condense to form aceto acetyl coA. Aceto acetyl CoA reacts with actyl CoA to form hydroxyl methyl glutaryl CoA (HMG CoA), which inturn gives rise to mevalonic acid. Mevalonic acid is phosphorylated three times and forms mevalonate pyro phosphate, which is converted to isopentenyl pyrophosphate, genanyl pyrophosphate and famesyl phyrophosphate. The two molecule of and famesyl phyrophosphate condense to form a hydrocarbon squalene. It undergoes ring closure and loss of methyl group and converted to cholesterol.

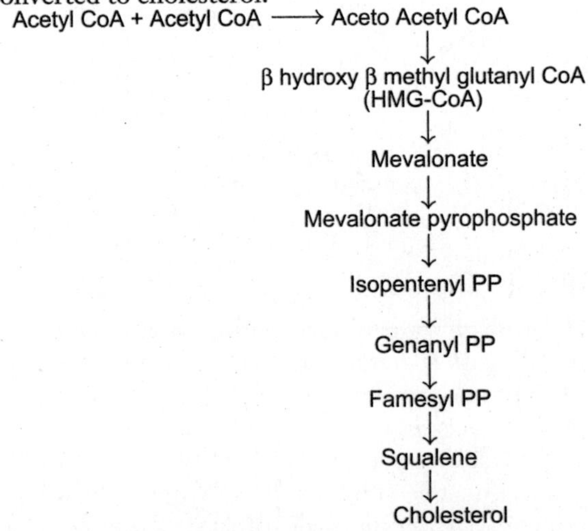

Acetyl CoA + Acetyl CoA $\longrightarrow$ Aceto Acetyl CoA

β hydroxy β methyl glutanyl CoA
(HMG-CoA)

Mevalonate

Mevalonate pyrophosphate

Isopentenyl PP

Genanyl PP

Famesyl PP

Squalene

Cholesterol

Fig. 121 : Biosynthesis of cholesterol.

Biosynthesis of lecithin/phospholipid

Phospholipids are synthesized from fatty acids and its precursors. Triose phosphate is reduced by the dihydroxy acetone phosphate to 3 glycerol phosphate, which is subsequently esterified by 2 fatty acid residues. The resulting diglyceride phosphatic acid is then activated by CTP to form CDP di glyceride. It undergoes transfer reactions with serine and a glycerophosphate releasing CMP. The reaction between CDP diglyceride and glycerol phosphate leads to the formation of phospholipid classes like phosphotidyl glycerol, lecithin and cardiolipin.

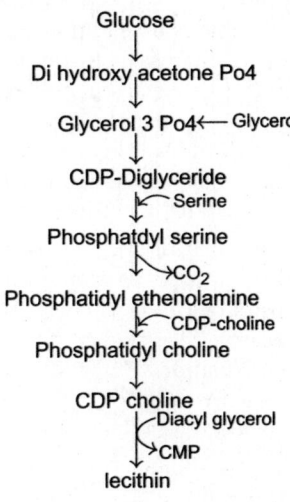

Glucose
↓
Di hydroxy acetone Po4
↓
Glycerol 3 Po4 ←— Glycerol
↓
CDP-Diglyceride
↓←— Serine
Phosphatdyl serine
↓→ CO_2
Phosphatidyl ethenolamine
↓←— CDP-choline
Phosphatidyl choline
↓
CDP choline
↓— Diacyl glycerol
↓→ CMP
lecithin

*Fig. 122 : **Biosynthesis of phospholipid.***

SPECTROPHOTOMETRY

Introduction

It deals with the study of the phenomenon of light absorption by molecules in solution. When a light at a particular wavelength is passed through a solution. Some amount of it is absorbed and thereby the light come out is diminished the nature of light absorption in a solution is governed by Beer – Lamberts law. Analytical procedures based upon the direct measurement of light absorption at specific wavelengths or regions of the spectrum are known as photometric procedures and the instruments used are photometers and spectrophotometers. In addition , there are methods which are dependent on the ability of insoluble particles to scatter light, called turbidometric methods and methods which are dependent on the ability of materials to emit light under specified conditions, called fluorimetric methods.

Spectrophotometric technique is based on the basic laws of light absorption. For uniform absorbing medium the proportion of the light radiation passing through it is called the transmittance, T, where $T = I/I0$. $I0$ = Intensity of the incident radiation, I= Intensity of the transmitted radiation. The extent of radiation absorption is more commonly referred to as the absorbance (A) or extinction (E) which are equal to the logarithm of the reciprocal of the transmittance,

$$i.e., A = E = \log 1/T = \log I0/I$$

Transmittance is generally expressed on a range of 0-100% and used in certain type of turbidity measurement. Absorbance or extinction varies from 0 to 100.

The Beer-Lambert Law

Lambert's law : it states that, when a ray of monochromatic light passes through an absorbing medium its intensity decreases exponentially as the length of the absorbing medium increases.

$$I = I0 \, e\text{-} k_1 l$$

Beer's law : it states that, when a monochromatic light passes through an absorbing medium its intensity decreases exponentially as the concentration of the absorbing medium increases.

$$I = I0 \, e\text{-}k_2 c$$

These two laws are combined together in the Beer- Lambert law:

$$I = I0 \, e\text{-}k_3 Cl$$

Colorimeter

White light from a tungsten lamp passes through a slit then a condenser lense, to give a parallel beam which falls on the solution under investigation contained in absorption cell or cuvette. The cell is made of glass with the sides facing the beam cut parallel to each other. In most of the colorimeters, the cells are 1 cm square and will hold 5 ml of solution. Beyond the absorption cell is the filter, which is selected to allow maximum transmission of the colour absorbed. If a blue solution is to be measured, a red filter should be selected. The colour of the filter is, therefore, complementary to the colour of the solution under investigation. In some instruments the filter is located before the absorption cell. The light then falls on to a photocell which generates an electrical current in direct proportion to the intensity of light falling on it. This small electrical signal is increased in strength by the amplifier , and the amplified signal passes to a galvanometer, or digital readout, which is calibrated with logarithmic scale and the extinction can be read directly. The blank solution (which does not contain the material under investigation) is first taken in the cuvette and reading adjusted to zero extinction and this is followed by the test solution and the extinction is recorded directly. A better method is to split the light beam, pass one part through the sample and the other through the blank, and balance the two circuits to give zero. The extinction is determined from the potentiometer reading which balances the circuit.

Procedure

There are four general steps in carrying out a photometric analysis:

(a) Separation of the substance from the complex mixture- for e.g., estimation of blood glucose requires the precipitation of lipids and proteins by using deproteinising agents which otherwise interfere with the colour reaction of glucose

(b) Quantitative conversion to a coloured or light absorbing substance- for e.g., after deproteinisation as mentioned above for glucose estimation, the supernatant is made to react with orthotoluidine reagent to give a greenish blue coloured complex

(c) Measurement of light absorption- for e.g., the colour intensity of the above mentioned complex is measured by using a red filter.

(d) Calculation of the concentration of the substance- for e.g., by comparing the extinction with that of the standard solution of the same substance of known concentration.

UV ABSORPTION SPECTROPHOTOMETRY

A spectrophotometer is a sophisticated type of colorimeter where monochromatic light is provided by a grating or prism in the place of filter in ordinary colorimeter. The band width of the light passed by a filter is quit broad, so that it may be difficult to distinguish between two compounds of closely related absorption with a colorimeter. Some compounds absorb strongly in the ultra violet region and their concentration can be determined by using a more expensive type of spectrophotometer which operates down to 190 nm. For e.g., (i) The activity of enzymes requiring NAD as coenzymes can be determined by treating the enzyme source with the relevant substrate and measuring the NADH formed (colourless) which gives strong absorption at 340 nm. The increase in absorbance is proportional to the concentration of the enzyme. (ii) the concentration of uric acid can be estimated by measuring the extinction of the solution at 293 nm before and after treatment with an excess of the enzyme uricase. At pH 9.0, uric acid which absorbs at 293 nm, is oxidized by uricase to allantoin, which has no absorption at this wave length. The decrease in absorbance at 293 nm is a measure of uric acid level.

Absorption spectra

Many compounds have characteristic absorption spectra in the ultra violet and visible regions so that identification of those materials in a mixture is possible.

Proteins : Proteins absorb strongly at 280 nm according to their content of the amino acids tyrosine and tryptophan, and this provides a sensitive and non-destructive form of assay.

Nucleic acids : Nucleic acids and their component bases show maximum absorption in the region of 260nm. The extent of absorption of nucleic acid is a measure of their integrity, since the partial degraded acids absorb more strongly than the native materials.

Haem proteins : These conjugated proteins absorb in the visible region as well as in the UV region of the spectrum due to haem group. The visible spectra of the oxidized and reduced forms of cytochrome C are sufficiently different so that the relative amounts of these forms can be determined in a mixture.

Applications of spectrophotometry

Colorimetry and spectrophotometry have widest application in biological sciences. These techniques are used for the determination of

(a) glucose, proteins, lipids, nucleic acid etc

(b) turbidity of solutions(bacterial cell mass)

(c) absorption spectrum of a compound

(d) purity of compound by knowing the molar extinction coefficient which is maximum for a pure compound.

CHROMATOGRAPHY

Introduction

The term chromatorgraphy was originally applied by Micheal Tswett, a Russian Botanist, in 1906. He defined chromatography as the process of separation and purification of coloured substances. Now a day it is used to identify, separate and purify one or more biological components in a mixture of a biological sample. Basically all chromatographic systems consists of two phases. One is the stationary phase which may be a solid, liquid or a solid liquid mixture which is immobilized. The mobile phase may be a liquid or a gas and flows over or through the stationary phase.

Types of chromatography

The interaction between stationary and mobile phase is often employed in the classification. Based on stationary phase chromatography are classified as column, Thin layer and paper chromatography.

1. **Column chromatography** : In this type, the stationary phase is packed into a glass or metal columns (wide tubes or cylinders). The mixture to be separated is layered on the top of the column in the form of a solution at particular concentration. After equilibration the components are eluted out of the column one by one using specific mobile phases (**Fig.123**). The solvent used to elute the separated components is known as eluant.

Column chromatography

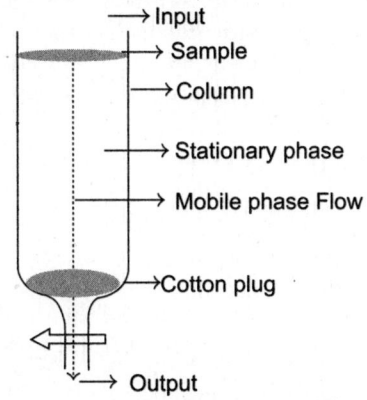

Fig. 123 : *Column chromatography.*

2. **Thin layer chromatography** : In this type, the stationary phase is thinly coated on to a glass, plastic or foil plates. The mixture to be separated is applied on the stationary phase at one end and kept vertical in the beaker containing the mobile phase. When the mobile phase reaches the other end of the plate the plate is removed from the beaker and the compounds separated are identified by using specific staining reagents.

3. **Paper chromatography** : In this type, the stationary phase is supported by the cellulose fibres of a paper sheet. The mobile phase flows through the stationary phase and effects separation.

Chromatography is again classified based on the usage of different mobile phases. They are as follows :

(a) partition chromatography   (b) adsorption chromatography
(c) ion-exchange chromatography   (d) exclusion chromatography
(e) affinity chromatography   (f) HPLC

(a) Partition Chromatography

This technique is based on the partitioning of compounds between a stationary phase and a liquid mobile phase. The liquid stationary phase can be held on any solid support like paper. It is of three types. They are paper chromatography, Thin layer chromatography and Gas – Liquid chromatography.

Paper chromatography : The cellulose fibres of chromatography paper act as the supporting matrix for the stationary phase. The stationary phase may be water or a non-polar material such as liquid paraffin. The components get separated between the liquid stationary phase and the liquid mobile phase. The procedure consists of paper development and component detection.

Paper development : There are two techniques which may be employed for the development of paper, ascending and descending methods. In both cases, the solvent is placed in the base of a sealed tank or glass jar to allow the chamber to become saturated with the solvent paper. The sample spots should be in a position just above the surface of the solvent so that as the solvent moves vertically up the paper by capillary action, separation of the sample is achieved.

Component detection : The separated components can be detected by (i) examining the paper under ultraviolet light; (ii) spraying of papers with specific colour reagents, for example ninhydrin for amino acids and sulphuric acid for simple sugars. The identification of a given compound may be made on the basis of its Rf value (retardation factor) which is the distance moved by the component during development divided by the distance moved by the solvent from the point of origin). The distance moved by the solute from origin.

$$Rf = \frac{\text{Distance travelled by solute}}{\text{Distance travelled by solvent front}}$$

The value of Rf is constant for a particular compound under standard conditions and closely reflects the distribution co-efficient for that compound.

Thin Layer chromatography

Partition, adsorption, exclusion chromatography can be carried out in a thin layer mode. In this technique the stationary phase is made in the form of a slurry and applied as a thin coating on the surface of a glass plate.After activating the plate, the sample to be separated is applied at one end of the plate . The plate is kept vertically in a chamber specially designed for this purpose(TLC chamber) and allowed the sample and the mobile phase to raise through the stationary phase by capillary action. The whole procedure consists of thin layer preparation, sample application, plate development and component detection.

(a) Thin layer preparation : A slurry of the stationary phase, generally applied to a glass, plastic or foil plate as a uniform thin layer by means of a plate spreader starting from one end of the plate and moving progressively to the other. Calcium phosphate is incorporated into the slurry in order to facilitate the adhesion of the adsorbent to the plate. The plate is heated in an oven at 1000 C to activate the adsorbent.

(b) **Sample application :** The sample is applied to the plate by means of a micropipette or syringe as spot or as a band on the stationary phase.

(c) **Plate development :** Separation takes place in a glass tank which contains mobile phase to a depth of about 1.5 cm. This is allowed to stand for atleast an hour with a lid over the top of the tank to ensure that the atmosphere within the tank becomes saturated with solvent vapour.

(d) **Component detection :** The components separated are detected by **(i)** spraying the plate with 50% sulphuric acid or 25% sulphuric acid in ethanol and heating; **(ii)** examining the plate under ultraviolet light; **(iii)** spraying of plates with specific colour reagents, for example ninhydrin for amino acids.

TLC

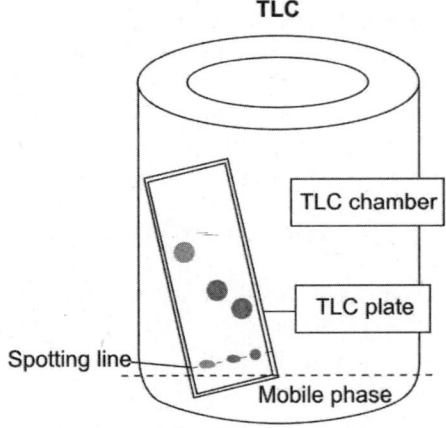

Fig. 124 : TLC.

(a) Gas liquid chromatography

This technique is based upon the partitioning of compounds between a liquid stationary phase and a gas mobile phase . It is a widely used method for the qualitative and quantitative analysis of a large number of compounds (eg. fatty acids) because it has high sensitivity, reproducibility and speed of resolution. A stationary phase of liquid material such as a silicone grease is supported on an inert granular solid . This material is packed into a narrow coiled glass or steel column 1 to 3 meter long and 2 to 4mm internal diameter. Through this column an inert carrier gas (the mobile phase) such as nitrogen, helium or argon is passed. The column is maintained in an oven at an elevated temperature which volatilizes the compounds to be separated.

The basis for the separation is the difference in the partition coefficients of the volatilized compounds between the liquid and gas phases as the compounds are carried through the column by the carrier gas. As the compounds flow, they leave the column and pass through a detector which is connected to a recorder and record a peak. The area of the peak corresponds to the concentration of the compound separated.

(b) Adsorption Chromatography

It is the oldest form of chromatography. It was developed by Tswett in 1903. The adsorbents used in this chromatography are finely divided porous solid particles, which can hold molecules on its surface, by electrostatic interactions. This chromatography can easily separate compounds, which are readily soluble in organic solvents. In this chromatography finely divided particles such as alumina, silicic acid, sugar, activated charcoal, cellulose are used as an adsorbent. Adsorption chromatography can be carried out in both the column and thin layer modes.

(c) Ion Exchange Chromatography

The principle of this form of chromatography is the attraction between oppositely charged particles. Many biological materials, such as amino acids and proteins, have ionisable groups and the fact that may carry a net positive or negative charge can be utilized in separating mixtures of such compounds. The net charge carried by such compounds depend on their pKa and on the pH of the solution. Ion exchange separations are mainly carried out in columns packed with an ion exchanger, which contain the core matrix molecule with exchangeable ionic groups on its surface. There are two types of ion exchangers, namely cation and anion exchangers. Cation exchangers posses negatively charged groups and they will attract positively charged molecules. Anion exchangers have positively charged groups which will attract negatively charged molecules. The actual ion exchange mechanism composed of four steps :

  (i) selective adsorption of the molecules to be separated by the ion exchange resins.

 (ii) release of the exchangeable group from the matrix.

(iii) Elution of the absorbed molecule by specific eluants.

(iv) Regeneration of the matrix by recharging with the original exchangeable groups.

Some of the ion exchange materials used in this technique are Amberlite IRC 50, Bio-Rex, Dowex 50, Sephadex etc.

MOLECULAR EXCLUSION CHROMATOGRAPHY

This chromatography is otherwise known as gel permeation chromatography. This technique is based on the separation of molecules on the basis of their molecular size and shape and the molecular sieve properties of a variety of porous materials which serve as the solid stationary phase. A column of gel particles or porous glass granules is in equilibrium with a suitable solvent for the molecules to be separated. Large molecules which are completely excluded from the pores will pass through the interstitial spaces and smaller molecules will be distributed between the solvent inside and outside the molecular sieve and will then pass through the column at a lower rate.

AFFINITY CHROMATOGRAPHY

This technique is based on the specific biological interaction of the compounds to be separated with the special molecules attached on the stationary phase called as ligands. This technique requires that the material to be isolated is capable of reversibly binding to a specific ligand which is attached to an insoluble matrix (stationary phase). Under suitable experimental conditions when a complex mixture containing the specific compound to be purified is added to the insolubilised ligand generally contained in a chromatography column, only that compound will bind to the ligand. All the other compounds can be washed away and the compound subsequently recovered by displacement from the ligand. In practice, particles which are uniform, spherical and rigid are used as matrix materials such as polystyrene, cellulose, porous glass and silica etc. it is used for the separation of enzymes, vitamins, nucleic acid etc.

HPLC

Generally chromatographies are done with slow speed. They are time consuming process. The separation can be improved by applying high pressure. It uses non compressive resin material and strong metal column. Elutants of the column are detected by UV absorption and fluorescence.

Applications of chromatography

1. Thin layer chromatography is used for the separation of alkaloids, phospholipids and other lipids.

2. Gas liquid chromatography is applied for the separation of fatty acids in a lipid mixture.

3. Ion exchange chromatography can be used for the separation and identification of amino acids in a mixture of protein hydrolysate. This principle is used in auto-analyzer.

4. Exclusion chromatography can be applied for the determination of the molecular weight of the components separated.

5. Affinity chromatography is applied for the purification of a wide range of enzymes and other proteins like immunoglobulins.

ELECTROPHORESIS
(Gel, Isoelectric focusing, Immunoelectrophoresis)

Introduction

Movement of charged particles in an electric field resulting in the migration towards the oppositely charged electrode is known as electrophoresis. Biological molecules such as amino acids, peptides, proteins, nucleotides and nucleic acids posses ionisable groups and can therefore be made to exists in solution as electrically charged either as cations (+) or anions (-). These components are subjected to electric field, they migrate differentially and thus can be separated. Gel electrophoresis – native poly acrylamide gel electrophoresis and sodium dodecyl sulphate (SDS) poly acrylamide gel electrophoresis, isoelectric focussing and immunoelectrophoresis are the some types of electrophoresis.

The most commonly used electrophoresis is gel electrophoresis. In this technique either agarose or poly acrylamide is used as supporting media. Electrophoresis units are available for running either vertical or horizontal gel systems.

Agarose Gels : Agarose is a linear polysaccharide made up of repeating units of agarobiose which contains galactose and 3, 6 anhydro galactose. This is isolated from seaweeds. Agaroses gel is usually prepared at the concentration of 1-3% solutions. The gels can be prepared by suspending dry agarose in suitable aqueous buffer then boiling the mixture until a clear solution forms. This is poured and allowed to cool to room temperature to form a rigid gel. Agarose gels are used for the electrophoresis of both proteins and nucleic acids.

Polyacrylamide Gels: Electrophoresis in acrylamide gels is referred as **PAGE** being an abbreviation for **P**oly**A**crylamide **G**el **E**lectrophoresis. Polyacrylamide gels are prepared by dissolving required quantity of acrylamide with a small amount of N, N'-methylene bisacrylamide in suitable buffer. The polymerization is initiated by ammonium persulphate and N, N, N', N'-tetramethylene diamine (TEMED).The polymerization is free radical mediated reaction. Sodium dodecyl sulphate (SDS) polyacrylamide gel electrophoresis is the most widely used method for analyzing protein mixture qualitatively. It is particularly useful for monitoring protein purification. SDS is an anionic detergent. Samples to be run on SDS-PAGE are first boiled for 5 minutes in sample buffer containing beta mercapitoethanol and SDS. The mercapto ethanol reduces any disulphide bridges and cleave the protein into different sub-units. So, by this electrophoresis different units of proteins can be identified.

Detection of separated components

1. Proteins can be detected by using the dye-solution Coomassie Brilliant Blue R-250 (CBB). Staining is usually carried out using 0.1% CBB in methanol : acetic acid : water in the ratio 5:1:5. The protein bands will look blue in colour.

2. Glycoproteins are detected by using periodic acid – schiff (PAS) stain. The bands will appear in red colour.

2. Nucleic acids can be detected by using the fluorescent dye ethidium bromide. The nucleic acids bands will appear colourless in black background.

Isoelectric focusing

It is also known as electrofocusing, is a technique for separating different molecules by differences in their isoelectric point (pI). It is a type of zone electrophoresis, usually performed on proteins in a gel, that takes advantage of the fact that overall charge on the molecule of interest is a function of the pH of its surroundings.

Immunoelectrophoresis

Immunoelectrophoresis is a powerful analytical technique with high resolving power as it combines separation of antigens by electrophoresis with immunodiffusion against an antiserum. The increased resolution is of benefit in the immunological examination of serum proteins. Immunoelectro-phoresis aids in the diagnosis and evaluation of the therapeutic response in many disease states affecting the immune system. It is usually requested when a different type of electrophoresis, called a serum protein electrophoresis.

CENTRIFUGATION

Introduction

These techniques are based up on the behaviour of particles in an applied centrifugal field such as density, shape or size of the molecules being separated. The particles are normally suspended in a specific liquid medium held in tubes or bottles which are located in rotor. The rotor is positioned centrally on the drive shaft of the centrifuge. Particles which differ in density, shape and size can be separated since they sediment at different rates in the centrifugal field. Each particle sediments at a rate which is proportional to the applied centrifugal field. The rate of sedimentation can be expressed as rpm (revolutions per minute) or as g (gravitational force). The rate of sedimentation is dependent up on the applied centrifugal field (G) which is determined by the square of the angular velocity of the rotor (ω) and the radial distance (r) of the particle from the axis of rotation according to the equation $G = \omega 2r$. The sedimentation rate or velocity (v) of a particle can also be expressed in terms of its sedimentation per unit centrifugal field, known as sedimentation co-efficient (s).

$$v = s\omega 2r.$$

Types of centrifugation techniques : Two main types of centrifugation techniques are in general use.

1. Preparative centrifugation techniques: This technique is applied for the actual separation, isolation and purification of whole cells, plasma membrane, ribosomes, chromatin, nucleic acids, lipoproteins, viruses and many sub-cellular organelles. Large amount of materials may be involved for the bulk isolation.

2. Analytical centrifugation techniques: This technique is applied to study the characteristic features of pure macromolecules or particles. It requires only a small amounts of materials and utilizes specially designed rotors and detector system to continuously monitor the process of sedimentation of the material in the centrifugal field.

Centrifuges and their uses : The instrument used for this technique is known as centrifuge.Four major types of centrifuges are generally used. They are

(a) small bench centrifuges
(b) large capacity refrigerated centrifuges

(c) high speed refrigerated centrifuges

(d) ultracentrifuge (i) preparative (ii) analytical

(a) **Small bench centrifuges** : These are the simplest and less expensive instruments. They are used to isolate erythrocytes from blood and other separations which require low centrifugal force. These centrifuges generally have a maximum speed of 4000-6000 rpm (revolutions per minute). The speed can also be expressed as g / min.

(b) **Large capacity refrigerated centrifuges** : These centrifuges have a maximum speed of 6000 rpm/ min . Compounds to be separated can be taken in bulk. The instrment is provided with refrigeration facility . By this method, biological materials can be isolated without any loss in their biological properties. Erythrocytes, coarse or bulky precipitates, yeast cells, nuclei and chloroplasts can be isolated by using this centrifuge.

(c) **High speed refrigerated centrifuges** : These instruments have maximum speed of 25000 rpm/min. They are used to collect microorganisms, cellular debris, large cellular organells and precipated proteins

(d) **(i) Preparative ultracentrifuges** : A maximum speed of 80000 rpm / min can be attained by this centrifuge. The rotor chamber is refrigerated, sealed and evacuated to minimize excessive rotor temperature. These centrifuges are used for the separation of lipoprotein fractions and for deproteinisation of physiological fluids for aminoacid analysis

(ii) **Analytical ultracentrifuges** : These instruments are capable of operating at about 70000rpm/min. The rotor is present inside an evacuated, refrigerated chamber. An optical system is attached to observe the materials getting sedimented and to determine concentration distributions within, at any time during centrifugation. This technique finds applications for the separation and isolation of hormones, enzymes , ribosomal units, viruses and subcellular organells from animal and plant tissue homogenates.

Differential centrifugation technique

It is a type of preparative centrifugation. This method is based on the differences in the sedimentation rate of particles of different size and density. In differential centrifugation, the material (a tissue homogenate) to be separated in solution is centrifugally divided in to a number of fractions by the step wise increase of applied centrifugal field. The centrifugal field is determined by trial and error method so that the particular type of material sediments during predetermined time of centrifugation to sediment the particles in the form of pellet. The supernatant contains other materials which are unsedimented. At the end of each stage the pellet and supernatant are separated and the pellet is purified by washing. Initially, all particles of the homogenate are homogenously distributed through out the centrifuge tube. During centrifugation, particles move down the centrifuge tube at their respective sedimentation rates and start to form pellet at the bottom of the tube. Centrifugation can be continued till all the components are pelleted one by one by increasing the centrifugal field. For example, the sub-cellular organelles (nucleus, mitochondria, lysosomes, microsomes) from a tissue liver homogenate can be isolated by applying this differential centrifugation techniques. The technique has the following steps:

1. Preparation of liver homogenate – 10% solution in 0.25 molar sucrose.

2. Centrifugation at 1000 g for 10 minutes.

3. Isolation of the pellet sedimented which is nucleus.

4. The supernatant decanted from step (c) is subjected to centrifugation at 3300 g for 10 minutes.

5. Isolation of the pellet sedimented which contains mitochondria.

6. The supernatant decanted from step (e) is subjected to centrifugation at 16300 g for 20 minutes.

7. Isolation of the pellet sedimented which contains lysosomes.

8. The supernatant decanted from step (g) is subjected to centrifugation at 105000 g for 60 minutes.

9. Isolation of the pellet sedimented which contains microsomes.

10. The supernatant obtained in the final step is the cell free cytosol.

The isolation of sub-cellular organelles is an essential procedure used in many biochemical research laboratories by using this differential centrifugation techniques.

Applications of analytical ultra-centrifuge : The analytical ultra-centrifuge has found many applications in fields of protein and nucleic acid chemistry. This gives information about (a) determination of molecular weight of biomolecules, (b) estimation of purity of macromolecules and (c) detection of conformational changes in macromolecules.

pH

Introduction

pH is defined as the negative logarithm of hydrogen ion concentration. **The term pH was introduced by Sorenson in 1909. It express as power of Hydrogen ion concentration. pH is defined as negative logarithm of hydrogen ion concentrations.**

i.e. $pH = -\log [H^+]$ (or) $pH = \log 1/[H^+]$

Acidic : 1 - 6.9 Neutral 7.0 Alkaline 7.1 - 14.

Determination of pH-Henderson - Hasselbalch equation Derivation

This concerns the dissociation of weak acid in equilibrium.

Let us consider HA a weak acid that ionises as follows :

$$HA \rightleftharpoons [H^+] + [A^-]$$

$$\therefore K_a = \frac{[H^+] + [A^-]}{[HA]} \text{ [where } K_a = \text{Dissociation constant]}$$

$$\therefore H^+ = \frac{Ka \times [HA]}{[A^-]}$$

log [H+] = log Ka + log[HA]-log[A-] (Taking log on both sides).

-log [H+] = -log Ka -log [HA] + [A-] [Chaning sign on both sides].

pH = pKa + log [A-] / [HA] since -log H+ = pH and - log Ka = pKa

The above equation is known as Henderson - Hasselbalch equation. and can be used for the determination of pH of blood.

Determination of pH of buffers

The pH buffers can be determined by Henderson-Hasselbalch equation

$$pH = pK_a + \log \frac{[Salt]}{[Acid]}$$

$$pH = pK_a + \log \frac{[Salt]}{[Acid]}$$

where

$$pk_a = \log 1/K_a, K_a = \text{Dissociation Constant of an acid and}$$

$$K_a = \frac{[H^+] [Ac^-]}{[HAC]} \text{ for the reaction}$$

$$HAC \rightleftharpoons [H^+] + [Ac^-]$$

$$\therefore H^+ = \frac{K_a \times [HA]}{[A]}$$

In case of blood, the ratio between $[BHCO_3]$: $[H_2CO_3]$ can be found out by applying the above equation to maintain average pH of blood 7.4. If the pk_a value of H_2CO_3 is 6.1 then

$$7.4 = 6.1 + \log \frac{[BHCO_3]}{[H_2CO_3]}$$

$$\therefore \log \frac{[BHCO_3]}{[H_2CO_3]} = 1.3$$

$$\therefore \log \frac{[BHCO_3]}{[H_2CO_3]} = \log 20$$

$$\therefore \log \frac{[BHCO_3]}{[H_2CO_3]} = 20/1$$

So, $[BHCO_3]$: $[H_2CO_3]$ = 20 : 1

To determine exactly how acidic or basic a solution is, you must measure the concentration of hydronium (H_3O^+) ions.

The pH of a solution is often critical.

Ex: Enzymes in your body work only in a narrow pH range.

pH values correspond to the concentration of H_3O^+ ions, but also indicates OH^- ion concentrations.

The pH scale ranges from 0 to 14.

- Water is neutral and has a pH of 7.
- Solutions with a pH of less than 7 are acidic.
- Solutions with a pH of greater than 7 are basic.
- The further from 7 you go the more acidic or basic something is

$$pH = -\log_{10}[H_3O^+(aq)]$$

Thus, the pH of neutral water at 25°C is calculated

$$pH = -\log(10^{-7}) = -(-7) = +7$$

and the pH of pure water is 7.

Calculate the pH of an acidic solution prepared by adding enough water to 5 grams of hydrochloric acid to make a solution that has a volume of 1 liter.

$$\frac{5g\ HCl/L}{36.46\ g/mole} = 0.137\ mole.$$

Assuming the hydrochloric acid is completely dissociated,

$$HCl\ (aq) \rightarrow H_3O^+\ (aq) + Cl^-\ (aq)$$

The concentration of the hydronium ions would be 0.137 M, and the pH of the solution equals

$$pH = -\log(0.137) = 0.86$$

Notice that this acidic pH is a positive number less than 7 because the calculation takes the *negative* logarithm of $[H_3O^+]$. Try this calculation now to make sure that you understand this point. The pH of a solution may be estimated from color indicators that change hue with pH, like litmus or phenolphthalein papers. If precise values are required, an electrical pH meter is utilized.

BUFFER

Introduction

A buffer is a mixture of a weak acid and its salt with a strong base (eg) A mixture of acetic acid and sodium acetate.

$$HAC + NaAC \longrightarrow Na^+ + H^+ + 2AC$$

Where   HAC = Acetic acid; NaAC = Sodium acetate.

A buffer solution is one which resists a change in its pH value (hydrogen ion concentration) on dilution or on addition of an acid or alkali. The process by which added H^+ and OH^- ions are removed so that pH remains constant is known as buffer action. For (eg) if alkali is added to the above mentioned buffer it forms NaAC and no free H^+ or OH^- will be available.

$$[Na^+ + H^+ + 2AC^-] + NaOH \longrightarrow 2NaAC + H_2O$$

If an acid is added to the buffer it will form NaCl and no free H^+ or OH^- will be available.

$$[Na^+ + H^+ + 2AC^-] + HCl \longrightarrow NaCl + 2HAC$$

In either cases there is no change in hydrogen ion concentration i.e. it resist the change in pH of the solution.

Examples of buffer

A mixture of

1. Glycine and HCl
2. Potassium dihydrogen phosphate and dipotassium hydrogen phosphate.
3. Sodium bicarbonate and sodium carbonate.

Uses of buffer

1. Bufers are used for preparing standard solutions in which it is always desired to maintain a constant pH.

2. These are used to maintain H+ concentration which is necessary for optimal activity of enzymes.

3. Buffers regulate acid-base balance by restricting pH change in body fluids and tissues, since they are capable of absorbing H^+ ions and OH^- ions when an acid or an alkali is formed during metabolic activities.

Bicarbonate buffer

It is the most important buffer in blood plasma and consist of bicarbonate $[HCO_3^-]$ and carbonic acid $[H_2CO_3]$. This buffer is efficient in maintaining the pH of blood plasma to 7.4 against the acids produced in tissue metabolism (eg) phosphoric acid, lactic acid, aceto acetic acid and β-hydroxy butyric acid. These acids are converted to their anions and the bicarbonate is converted to carbonic acid a weak acid.

$$HCO_3^- + H^+ \text{ (obtained from acids)} \longrightarrow H_2CO_3 \longrightarrow CO_2$$

CO_2 thus formed is expirated through lungs during respiration.

Phosphate buffer

The phosphate buffer consists of dibasic phosphate [HPO42-] and monobasic phosphate ($H_2PO_4^-$). Its pKa value is about 6.8. It is more effective in the pH range 5.8 to 7.8. Plasma has a ratio of 4 between

$$[HPO_4^{2-}] : [H_2PO_4^-].[HPO_4^{2-}].$$

Therefore

$$pH = pKa + \log \frac{[HPO_4^{2-}]}{[H_2PO_4^-]}$$

$$pH = 6.8 + \log 4 = 7.4$$

Protein buffer

The protein buffers are very important in the plasma and in the intracellular fluids but their concentration is very low in CSF, lymph and interstitial fluids.

They exist as anions serving as conjugate bases (Pr⁻) at the blood pH 7.4 and form conjugate acids (HPr) accepting H⁺. They have the capacity to buffer some H_2CO_3 in the blood.

$$H_2CO_3 + Pr \rightarrow HCO_3^- + HPr^{\bullet\bullet} \rightarrow H_2CO_3 \rightarrow CO_2\uparrow$$

Hemoglobin buffer

They are involved in buffering CO_2 inside erythrocytes. The buffering capacity of hemoglobin depends on its oxygenation and deoxygenation. Inside the erythrocytes, CO_2 combines with H_2O to form H_2CO_3 under the action of carbonic anhydrase. At the blood pH 7.4, H_2CO_3 dissociates into H+ and HCO_3^- and needs immediate buffering. Oxyhemoglobin (HbO_2^-) on the other side loses O_2 to form deoxyhemoglobin (Hb⁻) which remains undissociated (HHb) by accepting H+ from the ionization of H_2CO_3. Thus, Hb- buffers H_2CO_3 in erythrocytes.

$$HbO_2 \rightarrow Hb + O_2$$

$$Hb + H_2CO_3 \; HHb + HCO_3^-.$$

ACID AND BASES

Introduction

According to the modern concept, an acid is defined as that species which can donate H⁺ ions (protons) in solution and a base is that species which can accept H⁺ ions. Since such transfer of protons is reversible any acid which gives up its proton becomes a base, while any base which accepts a proton becomes an acid. This theory was postulated by Bronsted and Lowry in 1923. The acid is capable of forming a covalent bondwith an electron pair (a Lewis acid).

$$\text{Acid} \rightleftharpoons H^+ + \text{Base}$$

An acid and a base related in this manner are called conjugates.

For (eg) in this reaction

$$HCl \rightleftharpoons H^+ + Cl^-$$

HCl is acid and Cl⁻ is its conjugate base.

Other examples

$$H_2CO_3 \rightleftharpoons H^+ + HCO_3^-$$

$$H_2SO_4 \rightleftharpoons H^+ + HSO_4^{2-}$$

An acid which dissociates strongly and readily gives H+ ions is known as a strong acid. The capacity of an acid to release its protons is known as acidity. A base which has more affinity to combine with H+ ions is known as a strong base. This property of a base is known as alkalinity (eg) HCO_3^-, HSO_4^-, H_2PO_4 etc. The acidity of a species is denoted by its pH value : larger the acidity of a species, lower will be its pH value.

All acids ionize when dissolved in water. Strong acids ionize completely. When acids ionize water, charged ions are produced and they move around in the solution and are capable of conducting electricity.

1. A substance that can conduct electricity when dissolved in water is an **electrolyte**.

2. **Nitric acid**, HNO_3, is a strong acid and conducts electricity well.

3. Sulfuric acid, H_2SO_4, is also a strong acid and is used in car batteries to conduct the electricity needed.

4. Strong acids are make the strongest electrolytes because they form the maximum number of hydronium ions. They make the strongest electrolytes.

Weak acids do not ionize completely.

Acetic acid, CH_3COOH, does not conduct electricity well because it does not ionize completely. Therefore, it is a weak electrolyte.

Any acid can be dangerous in a concentrated form.

1. Acids are used in many manufacturing processes and are necessary to many organisms.

2. Hydrochloric acid, HCl, is a very strong acid. It can burn your skin but yet it is in your stomach to help you digest your food.

The number of acid molecules that combine with water to form ions and the ions that recombine to form acid molecules are in equilibrium.

Base : Any compound that increases the number of hydroxide ions when dissolved in water.

1. Bases have a bitter, soapy taste, and feel slippery.

2. Solutions of bases contain ions just like acids and can also conduct electricity.

3. Bases also cause indicators to change colors. Red litmus paper will turn blue.

Some bases contain hydroxide ions, OH^-, but others do not but they will react with water molecules to form OH^- ions. Strong bases are ionic compounds that contain a metal ion and a hydroxide ion. They are known as metal hydroxides. When a metal hydroxide is dissolved in water, the metal ions and the hydroxide ions dissociate, or separate.

1. Sodium hydroxide, NaOH, is a metal hydroxide found in some drain cleaners. It is also a strong electrolyte.

2. Some metal hydroxides, such as calcium hydroxide, $Ca(OH)_2$, and magnesium hydroxide, $Mg(OH)_2$, are not very soluble in water, but the ions that do dissolve separate completely. Calcium hydroxide is used to treat soil that is too acidic.

Like acids, bases can be very dangerous in concentrated form.

1. Sodium hydroxide, NaOH, and potassium hydroxide, KOH, can be dangerous even in fairly diluted form.

2. Bases are usually more dangerous than acids because they attack living tissue more rapidly.

When 2 moles of KOH is added to water, we get 2 moles of water molecules.

Weak bases ionize in water to form hydroxide ions. Ammonia, like other bases, forms hydroxide ions when it dissolves in water. In this process, water acts as an acid and donates a hydrogen ion to ammonia to form an ammonium ion, NH_4^+, and leaves a hydroxide ion, OH^-, behind. These bases are poor conductors of electricity. In this situation, NH_4^+ acts as a base and accepts a hydronium ion.

CHAPTER 16

QUESTION BANK

TWO MARK QUESTIONS'

Briefly write about the following

Glucose
Hexose
Carbohydrates
Pentose
Monosaccharides
Oligosaccharides
Polysachharides
Anomers
Polypeptides
Oligopeptides
Essential aminoacids
Non- essential amino acids
General structure of amino acids
Peptide bonds
Glycosidic bonds
Amino acids
Fatty acids
Essential fatty acids
Non essential fatty acids
Steroids
Formed elements
Plasma
Serum
WBC
RBC
Granulocytes
Coagulation
Buffy coat/
Endogrine glands
Homeostasis
STH/GH
Gigantism

Acromegaly
Hyperglycaemia
Chlorophyll
Carotenoids
Phycobilins
Hormones.
Ductless glands
Exocrine glands
Cholesterol
Phospholipids
Derived lipids
PUFA
DNA
RNA
Nucleic acids
Purines
Pyrimidines
Nucleosides
Nucleotides
mRNA
rRNA
Water soluble vitamins
Fat soluble vitamins
Vitamers
Vitamins
Retinol/Retinal/Retinoic acid
Beta carotene
Ergosterol
Bitot spots/Scurvy/Pellagra
Agranulocytes
Haematopoitic stem cells.
Anthocyanins
Accessary pigments
Phycocyanins
Chlorophyll a/b/ c
Phytohormones
Auxin a and b

Hetero auxin

Gibbane ring skeleton

Gibberellin A

Bolting and Parthenocarpy

Cytokinesis

Abscisin II

FIVE MARK QUESTIONS

Give a brief note on carbohydrates.

Bring out the functions of Monosaccharides.

State the biological significance of glucose.

Explain the structure of dextrose

How are carbohydrates digested? Describe.

Give a summary on the absorption of carbohydrates.

Describe the classification of monosccharides.

Discuss briefly the differences between homo and hetero polysaccharides.

What are essential amino acids? Explain with examples.

List out the non essential amino acids. Why are they referred so?

What are known as protein and non protein amino acids?

How would you estimate proteins?

Explain the quaternary structure of proteins.

Describe briefly the terms dipeptide, tripeptide, oligo peptide and polypeptides.

Explain the structure of amino acids.

Bring out the biological significance of proteins.

Explain the structure of lipids.

What are fatty acids? Explain.

Differentiate between simple and compound lipids.

Give a brief note on derived lipids.

Write the classification of fatty acids.

What are essential fatty acids? Write briefly.

Briefly mention about the properties of non essential fatty acids.

What is cholesterol? Explain.

Give differences between nucleotides and nucleosides.

Explain the biochemical composition of DNA.

What is the biochemical composition of RNA?

Discuss the functions of nucleic acids.

What is known as genetic RNA? Explain.

Explain the structure and functions of mRNA.

Give a note on the characteristic features of ribosomal RNA.

What are tRNAs? Explain their structure and functions.

Define vitamins. State the differences between fat and water soluble vitamins.

Give the classification of vitamins.

State the sources and deficiency diseases of vitamin D.

Write the sources and deficiency diseases of vitamins D & E.

How is vitamin K important? Explain.

Explain the deficiency diseases of vitamin E & K.

Describe the functions of vitamin A.

Write a summary on the deficiency diseases of vitamin C.

What are granulocytes? Explain.

Describe briefly agranulocytes.

What is coagulation? Explain.

How can we characterise WBC? Explain.

Explain the characterization of RBC.

State the normal ranges of various blood cells.

What are platelets? Write in brief.

Differentiate among haematocrit, buffy coat and serum.

Bring out the functions of insulin

Write briefly about adrenaline.

Describe about thyroid hormone.

Explain the endocrine glands – testis and ovary.

What is adrenal gland? Explain.

Give a note on parathyroid gland.

Write about pituitary gland.

What is known as PTH? Explain.

Write briefly about plant pigments.

What are accessory pigments? Explain their role.

List out the types and functions of chlorophyll pigments.

Give the biochemistry of carotenoid pigments.

Discuss briefly the functions of carotenoid pigments.

What are anthocyanin pigments? Explain.

Give the functions of phycobilins.

Write a general account on the importance of plant pigments.

What are phyto hormones? Explain.

Give a short account on secondary metabolites of plants.

Explain the biochemistry and structure of auxins.

List out the natural auxins.

Bring out the functions of gibberrlins.

Explain the term parthinocarpy and abscission.

Describe the structure of cytokinins.

Explain the structure of abcisic acids.

TEN MARK QUESTIONS

Elaborate the classification of carbohydrates.

Write in detail the characteristic features of Dextrose.

Discuss in detail, the functions of carhohydrates.

How do we estimate glucose present in a sample? Explain in detail.

Write a summary on the classification of amino acids.

Describe the terms protein & polypeptides in detail.

Explain with suitable examples, the differences between primary and secondary structures of protein.

Comment on the characteristic features of tertiary structure of protein.

Write the classification of lipids.

Elaborate the functions and biochemical significance of lipids.

How are lipids estimated? Explain.

Discuss the properties of lipids.

Explain Watson and Crick model of DNA.

Describe in detail the forms of DNA.

Write a detailed account the structure of RNA.

Write about the methodology involved in nucleic acid estimation.

Give a detailed summary on the sources, functions and deficiency syndromes of B1 & B2 vitamins.

Write in detail the sources, functions and deficiency syndromes of B_3, B_6 and B_{12} vitamins.

Define fat soluble vitamins. State and explain the sources, and deficiency syndromes of vitamin A.

What are the sources and functions of vitamin C? Write in detail.

Write in detail the composition of WBC.

How do RBC originate? Elaborate.

Explain the process of blood coagulation.

Discuss the origin of various blood cells.

Discuss the biological functions of animal hormones.

Write in detail about pituitary gland.

Give an account on thyroxine.

Explain the classification of hormones.

Write a detailed summary on chlorophyll pigments.

List out the plant pigments. How are they important for plant functioning?

Give a summary on accessory pigments.

What are the functions of carotenoid pigments? Elaborate.

List out and describe the functions of auxin.

Explain the structure of gibberlins.

What are the functions of abcisic acids? Elaborate.

Write in detail the functions of cytokinins.

A. QUALITATIVE ESTIMATION OF CARBOHYDRATES

Aim : To analyse the availability of carbohydrates.

Background information : Carbohydrates are distributed in living tissues. They are made from carbon, hydrogen and oxygen. Carbohydrates are divided into three main classes namely Monosaccharides, oligosaccharides/ disaccharides and polysaccharides. Carbohydrates are a source of energy. Monosaccharides are the simplest carbohydrates and are often called single sugars. e.g. fructose, glucose and galactose. Glucose is the most important carbohydrate fuel in human cells. Two glucose molecules react to form the dissacharide maltose. Starch and cellulose are polysaccharides made up of glucose units. Galactose molecules look very similar to glucose molecules.

Materials Required

Anthrone reagent : This reagent can be prepared by dissolving 200mg of anthrone reagent in 100ml of conc sulphuric acid.

Fehling's A solution : Dissolve 34.65g cupric sulphate in water and makeupto 500mL.

Fehling's B solution : Dissolve 125g Potassium hydroxide and 173g potassium sodium tartarate (Rochelle salt) in water and make upto 500mL.

Molisch's reagent : 5% a naphthal in alcohol, i.e., 5g of a naphthal dissolved in 100mL of ethanol.

Iodine solution : 0.005% in 3% KI, i.e., 3g of KI dissolved in 100mL water and then 5mg of iodine is dissolved.

Benedict's solution : 17.3g of sodium citrate and 10g of sodium carbonate aredissolved in 75mL of water. 1.73g of $CuSO_4.5H_2O$ is dissolved in 20mL of water. Mix the $CuSO_4$ solution with alkaline citrate with constant stirring, finally the whole volume is made up to 100mL with water.

Barfoed's reagent: 13.3g of copper acetate in 200mL of water and add 2mL of glacial acetic acid.

Concentrated HCl

Concentrated H_2SO_4

Osazone Reagent : It is prepared by adding 0.2 g of phenylhydrazine hydrochloride and 0.3 grams of sodium acetate.

Bial's Reagent : Dissolve 3g orcinol in 500mL concentrated HCl, add 2.5mL of a 10% solution of ferric chloride hexahydrate, and dilute to one liter with water; this is Approximately 6M HCl. The reagent is stable for months, but its yellow colour gradually darkens and some precipitate forms; this doesn't seem to affect its reactivity.

Seliwanoff's Reagent : Dissolve 1g resorcinol in 330mL concentrated HCl, dilute to one liter (approx. 4 M HCl final). This reagent seems to be stable for more than a year, though we usually make less than the recipe specifies.

Principle

Molisch's test : Con. H_2SO_4 dehydrates carbohydrates to form furfural and its derivatives. This product combines with sulphonated a naphthal to give purple colour.

Iodine test : Iodine forms a coloured absorption complex with polysaccharides due to the formation of micellae aggregate. Iodine will form a polysaccharide inclusion complex.

Benedict's test : Carbohydrates with a potential aldehyde or ketone group have reducing property when placed in an alkaline solution. Cupric ions present in the solution will be reduced to cuprous ion. This will give a red coloured precipitate. Moreover, this test is more specific for reducing sugars.

Barfoed' test : Barfoed's reagent is weakly acidic and it is only reduced by monosaccharides. Prolonged boiling may hydrolyze the disaccharide to give false positive test.

Bial's test : When pentose is heated with con.HCl, furfural, which condenses with orcinol in the prescence of ferric ion to give a blue green colour.

Seliwanoff's test : Ketoses are dehydrated more rapidly than aldose to give a furfural derivatives, which then condenses with resorcinol to form a red colour complex.

Osazone test : Compounds containing aldehyde and keto groups form crystalline osazone with phenyl hydrazine hydrochloride. Osazone crystals have characteristic shape and melting point which helps in the identification of reducing sugar.

Procedure

Perform the following qualitative tests using unknown carbohydrates.

S. No	Experiment Name and Procedure	Observation	Inference
1.	Test For Solubility Water Acid Alkali Alcohol	Mono and di saccharides are highly soluble in water	This confirms the presence of carbohydrates other than polysaccharides.
		Polysaccharides are soluble in acid or alkali.	This confirms the presence of polysaccharides.
2.	**Molisch's Test** Add 2 drops of Molisch reagent to 2mL of the sugar solution and mix thoroughly and pour 5mL concentrated sulphuric acid along the sides of the test tube.	A purple ring was formed at the junction of two layers, which spreads on standing.	This shows the presence of carbohydrate.

S. No	Experiment Name and Procedure	Observation	Inference
3.	**Anthrone Test** To 2mL of test solution add 1mL Anthrone reagent.	Blue green complex formed	This shows the presence of aldehyde and ketone group.
4.	**Benedict's Test** To 8 drops of test solution add 5mL of Benedict's solution and heated to boiling.	Orange red precipitate obtained.	Presence of reducing sugar.
		No characteristic colour change.	Absence of reducing sugar.
5.	**Fehling's Test** To 5mL of test solution add equal volume of Fehling's A and B solution and heated to boiling.	A reddish brown precipitate formed	This shows the presence of reducing sugar.
6.	**Barfoed's Test** To 5mL of test solution add 5mL	Brick red precipitate is Obtained at the bottom of test tube.	Presence of reducing monosaccharide
		No characteristic colour change.	Absence of reducing monosaccharide.
7.	**Selivanoff's Test** To 2mL of Selivanoff's reagent add 3 drops of test solution and heated to boiling.	Cherry red colour Obtained.	Presence of fructose.
		No characteristic colour change.	Absence of fructose.
8.	**Iodine Test** Add 2 drops of iodine to 1mL of the test solution.	Deep blue colour	Presence of polysaccharide
		Dark brown colour	Presence of polysaccharide (Glycogen).
		No characteristic colour change.	Absence of polysaccharide.

S. No	Experiment Name and Procedure	Observation	Inference
9.	**Osazone Test** Mix 0.1 g sugar sample with 0.2 g phenyl hydrazine hydrochloride 0.3 g sodium acetate and acidified with 5 drops of glacial acetic acid and heated in a boiling water bath for 15 minutes.	Yellow colour precipitate formed. The following crystals may be observed under microscope (a) Needle shaped crystals (b) Needle shaped crystals (c) Palm leaf shaped crystals (d) Powder buff shaped crystals (e) Flower petal shaped crystals	This shows the presence of aldehyde and ketone group. Presence of Glucose Presence of Fructose Presence of Galactose Presence of Lactose Presence of Maltose is confirmed.
10.	**Bials Test** To 5mL Bials reagent add 2-3 mL of test solution and worm gently and cool under tap water.	Blue green colour obtained No characteristic colour change.	Presence of pentose sugar. Absence of pentose sugar.

Result

The given sample contains ————— carbohydrate.

B. QUALITATIVE TEST FOR PROTEINS

Aim : To identify the protein present in the given sample solution

Background information : Proteins are about 50% of the dry weight of most cells and are the most structurally complex macromolecules. Each type of protein has its own unique structure and function. Proteins are polymers of about 20 amino acids (the monomer). Amino acids are built from a central carbon bonded to four *different* groups. They are hydrogen (–H), amino group (–NH$_2$), carboxyl group (–COOH), and some side chain symbolized by "R". To form protein, the amino acids are linked by dehydration synthesis to form peptide bonds. The chain of amino acids is also known as a polypeptide. Some proteins contain only one polypeptide chain while others, such as hemoglobin, contain several polypeptide chains all twisted together.

Principle

Ninhydrin Test: Ninhydrin is a powerful oxidising agent reacts with α-amino-acids,between pH 4-8 to give a purple colour complex. Ninhydrin reagent is reduced to hydrindantin during reaction with α-amino-acids. The amino acid in turn is converted into an aldehyde. Ammonia and Carbon dioxide are evolved. Hydrindantin and ammonia interact with another molecule ofninhydrin to form Ruhemann's purple coloured complex.

Xanthoproteic Test : Amino acid containing aromatic chains will form Xanthoproteic acid when it is treated with Con. HNO_3 salts of these derivatives are orange in colour when treated with alkali.

Pauly's Test : Diazotised sulphanilic acid couples with amino phenol and immidazole to form a coloured azo compound in cold condition.

Millon's Test : Phenolic amino acid on treatment with Millon's reagent gives red colour. Mercuric sulphate forms a coloured compound with hydroxyl group of amino acid (Tyrosine).

Morner's Test : Amino acid containing aromatic hydroxyl group reacts with this reagent to give green colour. This test is to specify for amino acid containing aromatic hydroxyl group.(tyrosine)

Folin's Test : Amino acid containing aromatic ring reacts with this reagent to give blue colour.

Hopkin's Cole Test : This reaction is answered by tryptophan. This reaction is due to the presence of indole group in tryptophan.

Ehrlisch Test : Indole group containing amino acid reacts with this reagent to give purple colour or pinkish red coloured complex.

Sakaguchi Test : This reaction is specific for guanidino group of Arginine or protein containing Arginine.

Sulphur Test : Amino acids containing the thiol or sulphydryl group reacts with sodium plumbate to form a dark grey or black precipitate which is insoluble in dil.HCl.

Sodium Nitroprusside (Bollin's) Test : Amino acids containing the free thiol group (S–H) (due to cysteine) yield a red colour, with sodium nitroprusside in an ammoniacal environment. Cystine which contains disulphide linkage (S–S) may be reduced to cysteine using reducing agent such as sodium cyanide, sodium brohydride or sodium bisulphate which then yields a positive result.

Procedure

S. No	Experiment Name and Procedure	Observation	Inference
1.	**Solubility :**		
	(a) Water	Colloidal solution formed.	Presence of protein.
	(b) Dilute NaOH solution	Partially soluble.	Presence of protein.

S. No	Experiment Name and Procedure	Observation	Inference
2.	**Precipitation by neutral salt solution :** To 1mL of the substance add equal volume of a saturated solution of Ammonium sulphate (half-saturation).	White precipitate formed.	Presence of Globulin.
	The solution is saturated with ammonium sulphate (full saturation).	White precipitate is formed.	Presence of Albumin.
3.	**Precipitation by heavy Metals** To 1mL of the sample add equal volume of 5% Mercuric nitrate.	White precipitate formed.	Presence of protein.
4.	**Precipitation by alcohol** To 1mL of the substance add equal volume of alcohol.	White precipitate formed.	Presence of protein.
5.	**Heat coagulation** About 1mL of the substance is taken in clean test tube and heated.	Cloudy white precipitate formed by coagulation.	Presence of protein.
6.	**Biuret test** To 1mL of the substance add few drops of Biuret reagent.	Purple colour formed.	Presence of protein.
7.	**Ninhydrin test** To 1mL of the test solution add few drops of Ninhydrin reagent and heat for 2 minutes.	A purple colour obtained.	Presence of amino acid in the protein.
8.	**Xanthoprotic test** To 1mL of test solution add few drops of conc. nitric acid and heat it. Cool it. Then add few drops of 40% NaOH.	Yellow colour formed after the addition of conc. Nitric acid and this turns red on the addition of NaOH.	Presence of aromatic Amino acid in theProtein.

S. No	Experiment Name and Procedure	Observation	Inference
9.	**Pauly's test** To 1mL of test solution add few drops of 1% sulphanilic acid and Cool it in an ice bath. Then add 1mL of 5% NaNO$_2$ and 1mL of 1% Na$_2$CO$_3$.	Deep blue colour dye is obtained.	Presence of Tyrosine and histidine units in protein.
10.	**Millon's test** To 1mL of test solution add Millon's reagent and heat it for few minutes.	Red colour obtained.	Presence of phenolicgroup containing aminoacid (tyrosine) in the protein.
11.	**Morner's test** To 1mL of the test solution add Millons Reagent and heat it in a boiling water bath.	Green colour obtained.	Presence of Tyrosine in the protein.
12.	**Folin's phenol test** To 1mL of the test solution add equal volume of Folin's reagent followed by the addition of 1% of Na$_2$CO$_3$.	Blue colour obtained.	Presence of Tyrosine in the protein.
13.	**Aldehyde test** To 1mL of the test solution add 2-3 drops of 1% HCHO and then carefully add few drops of conc. H$_2$SO$_4$ along the sides of the test tube.	A violet colour ring formed at the junction of the two layers.	Presence of Tryptophanin the protein.
14.	**Ehrlisch's test** To 1mL of the test solution add few drops of Ehrlisch's reagent and heat it in a boiling water bath.	Pinkish red colour obtained.	Presence of Tryptophan in the protein.

S. No	Experiment Name and Procedure	Observation	Inference
15.	**Sakaguchi's test** To 2mL of test solution add few drops of alpha-naphthol in alcohol followed by the addition of 1mL of 20% NaOH and a few drops of Bromine water.	Red colour obtained.	Presence of Arginine in the protein.
16.	**Sulphur test** To 1mL of test solution add few drops of 45% NaOH and boil it for 2 minutes cool it, then add lead acetate.	Dirty black precipitate obtained.	Presence of Cysteine in the protein.
17.	**Sodium nitro-prusside test** To 2mL of test solution add few drops of 20% NaOH . Then add 1mL of sodium–nitroprusside followed by the addition of 1.5mL of 1% Glycine. Boil it for few minutes. Then add 1mL of 6N HCl.	Reddish purple colour obtained.	Presence of Methionine in the protein.
18.	**Molisch's test** To 1mL of the substance add few drops of Molisch's reagent then add Conc.sulphuric acid along the sides of the test tubes.	Violet colour ring formed at the junction of two layers.	Presence of Carbohydrate unit in the protein.

Results

The given protein (egg protein) contains the following, Arginine, Tyrosine, Tryptophan, Cysteine, Methionine and Carbohydrate.

C. QUALITATIVE TEST FOR AMINO ACIDS

Aim : To identify the amino acid present in the given sample solution

Principle : Amino acids are basic units of proteins. There are 21 amino acids, which occur, commonly in biological systems. Their reaction varies with the nature of -R- groups. In the Ninhydrin test amino acids react with mild oxidizing agents (Ninhydrin) at 70°C to form NH_3, CO_2 and aldehyde of amino acids. In the second step reduced Ninhydrin reacts with oxidized Ninhydrin in the presence of ammonia (NH_3) forming a blue-coloured products. There are specific test for aromatic amino acids and sulphur containing amino acids.

Materials required : Millon's Reagent: Dissolve 15g of Mercuric Sulphate in 100mL of 15% of Sulphuric acid.

Sulphanilic Acid : 1g of Sulphanilic Acid in 100mL of 10% HCl (1% solution).

Sodium Nitrate(5%) : 5g of $NaNO_2$ is dissolved in 100mL of water.

Sodium Carbonate : (1%)-Dissolve 1g of Na_2CO_3 in 100mL of water (1% solution)

Ehrlisch's Reagent : Dissolve 10g of p-dimethyl amino benzaldehyde in 100mL of 10% HCl .

α-Naphthol: 1g of α-Naphthol is dissolved in 100mL of alcohol.

Sodium hydroxide (40%): 40g of NaOH is dissolved in 100mL of water (40% solution).

Bromine Water: Few drops of Bromine in 100mL of water.

Lead Acetate(1%): 1g of lead dissolved in 100mL of water.

Sodium hydroxide (1%): 1g of NaOH dissolved in 100 mL of water. (1% solution).

Con.Sulphuric acid

Glacial acetic acid

Procedure

S. No	Experiment Name and Procedure	Observation	Inference
1.	**Ninhydrin Test** To 2mL of the test solution add 2mL of Ninhydrin reagent and boil for 5 minutes.	Blue/Purple colour formation	Presence of Amino acids
2.	**Xanthoprotein Test** To 2mL of the test solution add few drops of concentrated Nitric acid and then add a few drops of dilute sodium hydroxide	Yellow Colour formed after the addition of con. HNO_3 and this turns to red colour while NaOH is added No characteristic colour change.	Presence of Aromatic amino acid. Absence of aromatic amino acid.

S. No	Experiment Name and Procedure	Observation	Inference
3.	**Millon's Test** To few drops of the test solution add a few drops Mercuric Sulphate followed by the addition of Sodium nitrite. Then add a few drops of 5% $NaNO_2$.	Red colour is obtained No characteristic red colour.	Presence of phenolic group containing amino acid. Presence of Tyrosine Absence of phenolic group containing amino acid.
4.	**Hopkin's-Cole Test** To 2mL of Glyoxylic acid add 2mL of test solution and mixed well. Then carefully add 2mL of concentrated sulphuric acid through the sides of the test tube	Ring is formed at the junction of two layer	Presence of Tryptophan
5.	**Pauly's Test** Mix 1mL of sulphanilic acid to 2mL of the test solution and cool it in ice. Add 1mL of 5% sodium nitrite solution. After 5minutes add 5mL of 1% sodium carbonate solution	Deep red colour dye is obtained No characteristic coloured dye	Presence of Histidine and Tyrosine. Absence of Histidine, Tyrosine and Tryptophan.
6.	**Morner's Test** To few drops of the test solution add a few drops of Morner's reagent and heat the solution in a boiling water bath.	Green colour is formed. No characteristic colour change.	Presence of Tyrosine. Absence of Tyrosine.
8.	**Folin's Phenol Test** To 2mL of the test solution add equal 1mL of Folin's Phenol reagent and then add 1% Na_2CO_3	Blue colour is obtained. No characteristic colour change.	Presence of Tyrosine. Absence of Tyrosine.

S. No	Experiment Name and Procedure	Observation	Inference
9.	**Aldehyde Test** To few drops of the test solution add 1mL of 1% HCHO and then add 1mL of con.H_2SO_4.	A violet colour ring is formed at the junction of two layers. No characteristic colour change.	Presence of Tryptophan. Absence of Tryptophan.
10.	**Hopkin's Cole Test** To few drops of the test solution add 1mL of glyoxallic acid and then add con. H_2SO_4 carefully along the sides of the test tube.	A violet colour ring formed at the junction of two layers. No characteristic colour change.	Presence of Tryptophan. Absence of Tryptophan.
11.	**Ehrisch's Test** To 2mL of the test solution add equal volumes of ehrlish's reagent and heat the solution in a boiling water bath for a few minutes.	Pinkish red colour obtained. No characteristic colour change	Presence of Tryptophan. Absence of Tryptophan.
12.	**Sakaguchi's Test** 3mL of test solution is mixed with 1mL of 40% sodium hydroxide solution and add 2 drops of–napthol then a few drops of bromine water is added.	Red colour obtained No characteristic colour change	Presence of Arginine. Absence of Arginine.
13.	**Sodium Nitro-prusside Test** To 2mL of test solution and 1mL of 20% NaOH followed by the addition of sodium nitroprusside and 1mL of glycine. Now heat the mixture. Cool it. Then add 6N HCl slowly in drops through the sides of test tube.	Redddish purple colour is obtained. No characteristic colour change.	Presence of Methionine. Absence of Methionine.

S. No	Experiment Name and Procedure	Observation	Inference
14.	**Sulphur Test** To few drops of test solution add equal volumes of 45% NaOH and then heat for 2 min in boiling water bath. Cool it. Then add 5mL of lead acetate solution.	Dirty coloured black precipitate obtained. No characteristic change	Presence of Cysteine. Absence of Cysteine.

Result

The given sample contain ————— aminoacid.

D. QUALITATIVE ESTIMATION FOR LIPIDS

Aim : To analyze the presence of lipids.

Background information

Lipids are a broad group of naturally-occurring molecules which includes fats, oils, waxes, phospholipids, steroids (like cholesterol), and some other related compounds. Fats and oils are made from two kinds of molecules, which include glycerol and three fatty acids joined by dehydration synthesis. Since there are three fatty acids attached, these are known as triglycerides. The terms saturated, mono-unsaturated, and poly-unsaturated refer to the number of hydrogens attached to the hydrocarbon tails of the fatty acids as compared to the number of double bonds between carbon atoms in the tail. Triglycerides contain the maximum possible amount of hydrogens, these would be called saturated fats. The hydrocarbon chains in these fatty acids are, thus, fairly straight and can pack closely together, making these fats solid at room temperature. Oils, mostly from plant sources, have some double bonds between some of the carbons in the hydrocarbon tail, causing bends or "kinks" in the shape of the molecules. Because some of the carbons share double bonds, they're not bonded to as many hydrogens as they could if they weren't double bonded to each other. Therefore these oils are called unsaturated fats. Because of the kinks in the hydrocarbon tails, unsaturated fats can't pack as closely together, making them liquid at room temperature.

Materials Required

Ethanol, Chloroform, bi salts solution and detergents, Potassium bi sulfate, 10% alcoholic sodium hydroxide, Alcoholic bromine solution, Acetic anhydride, Sulphuric acid.

Procedure

S. No	Experiment Name and Procedure	Observation	Inference
1.	**Solubility Test** the substance is conducted with the following solvents		
(a)	Water	Insoluble	Presence of Lipids.
(b)	Sodium Hydroxide	Insoluble	Presence of Lipids.
(c)	Alcohol	Insoluble	Presence of Lipids.
(d)	Benzene	Soluble	Presence of Lipids.
(e)	Chloroform	Soluble	Presence of Lipids.
(f)	Ether	Soluble	Presence of Lipids.
2.	**Emulsification Test** The sample was emulsified with 5mL of water 5mL of bile salt solution and detergents	Temporary emulsion on vigorous shaking Highly stable emulsion	The water do not have the tendency to reduce the surface tension of water but the bi salts and detergents can break the large fat into small droplets and can reduce the surface tension of oil gently
3.	**Acrolein Test** To a pinch of potassium bi sulfate in a dry test tube added 4 drops of sample and heated	Pungent smelling fumes	Acrolein is evolved which has the pungent smell. All fats answer this test
4.	**Saponification Test** To 5mL of the sample added 2.5mL of ethanol and 10mL of 10% alcoholic sodium hydroxide. It was shaken well and placed in the boiling water bath for 15 minutes. Then it was made up to 20mL with water and divided into four equal parts and added Concentrated Hydrochloric acid Standard Saline 3 drops of calcium chloride solution 3 drops of magnesium chloride solution	A white precipitate Pale white layer on the surface A white precipitate.	The insoluble fatty acids are precipitated Sodium salts of fatty acids was salted out Calcium salt of fatty acid was salted out. Magnesium salt of fatty acid was salted out.

S. No	Experiment Name and Procedure	Observation	Inference
5.	**Test for Un saturation** 3drops of the sample and 3mL of ethanol were mixed well. Then added alcoholic bromine solution, until bromine imparts its colour	Colourless at first and gradually turned into yellow	Bromine reacts with the fatty acids and produce Di bromide which was colourless But when all the double bond are saturated with bromine further addition of bromine imparts the colour
6.	**Test for Cholesterol Libermann-Burchard Test** 2mL of sample added 2mL of chloroform and 10drops of acetic anhydride mixed well and the concentrated sulphuric acid was added along the sides of the test tubes	Deep green colour	This indicates the presence of Cholesterol.
7.	**Salkowski Test** To 2mL of sample add 2mL of chloroform then added equal volume of concentrated sulfuric acid.	Two layers appeared brown layer and low yellow fluorescent layer.	This confirms the presence of Cholesterol.

Result

The given sample contain ———————— lipid.

E. ESTIMATION OF PROTEIN BY BIURET METHOD

Aim : To perform the estimation of protein by Biuret method.

Principle : Biuret method is the simplest method for protein estimation. This method is sensitive to the amino acid composition of the protein. Its sensitivity is moderately constant from protein to protein and because of its simple procedure and quick result, it is used to estimate protein in crude extract over a large range of concentration. This method can also be used to monitor the concentration of protein during purification. This assay is based on copper ions binding to peptide bonds of protein under alkaline conditions to give a violet or purple colour. The intensity of the charge transfer absorption bond resulting from the Cu-protein complex is linearly proportional to the mass of protein present in the solution. The chromophore or light-absorbing center seems to be a complex between the peptide backbone and cupric ions.

Materials Required

Biuret reagent : Dissolve 1.5gm of CuSO4 and 4.5gms of Na-K tartarate in 250mL 0.2 N NaOH solution. Add 2.5gms of KI and make up the volume to 500mL with 0.2 N NaOH. 0.2 N NaOH.

Protein standard : Bovine serum albumin at a concentration of 1mg/mL in distilled water is used as a stock solution.

Procedure

- Pipette out standard protein solution into a series of tubes (0.0, 0.2, ..., 1mL) and make up the total volume to 4mL by adding water.
- The blank tube will have only 4mL of water.
- Add 6mL of biuret reagent to each tube and mix well.
- Keep the tubes at 37°C for 10 minutes during which a purple colour will develop.
- Measure the optical density of each tube at 520nm (green filter).
- Draw the standard graph to the known concentration of a protein and calculate unknown / test sample protein concentrations.

Result

The given sample contains .... mG of protein.

F. ESTIMATION OF PROTEINS BY BRADFORD METHODS

Aim : To estimate concentration of protein by Bradford method

Background information : A simple procedure for the determination of protein concentration in solutions is the Bradford protein assay which was described first by Bradford (Bradford et al., 1976). An estimation of protein concentration is essential to be done rapidly and accurately in many fields of protein study. The Bradford assay has become the preferred method for quantifying protein in many laboratories. This technique is simpler, faster, and more sensitive than the Lowry method. Furthermore, when compared with the Lowry method, it is subject to less interference by common reagents and nonprotein components of biological samples. The Bradford assay relies on the binding of the dye Coomassie Blue G-250 to protein. The quantity of protein can be estimated by determining the amount of dye in the blue ionic form. This is usually achieved by measuring the absorbance of the solution at 595 nm. The dye appears to bind most readily to arginyl and lysyl residues of proteins.

Materials required

Bradford reagent : The assay reagent is made by dissolving 100mg of Coomassie Brillient Blue G250 in 50 mL of 95% ethanol. The solution is then mixed with 100mL of 85% phosphoric acid and made up to 1 L with distilled water. The reagent should be filtered through Whatman no. 1 filter paper and then stored in an amber bottle at room temperature.

Protein standard : Bovine serum albumin at a concentration of 1 mg/mL in distilled water is used as a stock solution.

Procedure

Pipette 0.1mL to 1mL of protein standard in a series of test tubes, which contains 10 to 100 μg of protein and made up to 1mL using distilled water.

Use one mL of distilled water as the reagent blank.

Take 0.1 mL of unknown sample and made up to 1mL using distilled water.

Add 5 mL of bradford reagent to each tube and mix well by inversion or gentle vortex mixing. Avoid foaming.

Measure the Absorbance at 595nm of the samples and standards against the reagent blank between 2 min and 1 h after mixing.

Prepare standard curve using standard values.

Result

The given sample contains .... mG/mL of protein.

G. ESTIMATION OF CARBOHYDRATE BY ANTHRONE METHOD

Aim : To estimate the amount of carbohydrate present in the given sample

Principle : This method is used to estimate total carbohydrate present in the sample. The anthrone reaction is the basis of rapid and convenient method for the determination of hexoses and pentoses, either free or present in polysaccharides. Carbohydrates are dehydrated by concentrated sulphuric acid to form furfural (Hexose) or 5–hydroxymethyl furfurol (Pentose). Furfural or hydoxy methyl furfurol condenses with anthrone gives rise to a green coloured complex which was measured colourimetrically at 620-640nm.

Materials Required

Anthrone reagent (0.2%) : Dissolve 0.2g of anthrone in 5 mL of ethanol. Add slowly 75% of sulphuric acid till the mark reaches 100mL in standard measuring flask.

Stock standard (1000μg/mL) : Dissolve 100mG of glucose in 100mL of distilled water.

Working standard : Makeup 10 mL of stock standard to 100mL. It gives 100μg/mL concentration.

Unknown sample

Other materials : Spectrophoto meter, Aluminium foil, Water bath, Test tubes, Standard measuring flask, Cuvette, Micropipette, Pipette.

Procedure

Standard graph preparation

1. Prepare various concentration of the working standard solution in a series of test tube from 0.1mL to 1mL (10μg to 100μg).

2. Make up the volume to 1mL with distilled water.

3. Keep the tubes in an ice bath and slowly add 5 mL of the cold anthrone reagent and mix properly.

4. Close the tubes with aluminium foil and place it in a boiling water bath for 10 min

5. Cool the tubes and measure OD at 620 nm.

6. Blank should be prepared as per previous steps without adding test or standard solution.

7. Plot the graph and calculate the carbohydrate content of the sample given.

Testing unknown solution

1mL of test solution is taken in test tube.

Follow steps 2 to 6.

Calculate concentration of carbohydrate using standard graph.

Estimation of Carbohydrate

S. No	Volume of working standard in mL	Volume of water in (mL)	Concentration of working sample (μg /mL)	Volume of anthrone in mL	OD at 620nm
1	0.1	0.9	10	5	
2	0.2	0.8	20	5	
3	0.3	0.7	30	5	
4	0.4	0.6	40	5	
5	0.5	0.5	50	5	
6	0.6	0.4	60	5	
7	0.7	0.3	70	5	
8	0.8	0.2	80	5	
9	0.9	0.1	90	5	
10	1.0	0.0	100	5	

Test Sample 1mL

Observation

Green colour formation was noted and measured OD at 620nm

Result

The given test sample contains ——————————mG of glucose per 100mL.

H. ESTIMATION OF PROTEIN –LOWRY'S METHOD

Aim : To estimate the amount of protein present in the given sample.

Principle : Protein reacts with the Folin- Ciocalteu reagent to give a coloured complex. Tyrosine and tryptophan residues of protein reduce sodium tungstate and sodium molybdate anions in folin reagent which when combines with Copper of copper sulphate gives blue coloured complex (Hetero polymolybdenum and tungsten blue). The copper atom present in copper sulphate complexed with nitrogen atom of the peptide bond of protein during reaction time that is also a reason for blue/purple colour formation. The intensity of colour depends on the amount of these aromatic amino acids present and will thus vary for different protein.

Materials Required

Solution1 : Alkaline sodium carbonate solution - Take 2 g of sodium hydroxide in 400mL of double distilled water. Mix well and then add 10g of anhydrous sodium carbonate. Shake well and make up to 500mL using distilled water.

Solution 2 : Copper sulphate Solution - Dissolve 1 g of copper sulphate in 50mL of distilled water

Solution 3 : Sodium potassium tararate solution - Dissolve 1g of sodium potassium tartarate in 50 mL of water

Solution 4: Mixed reagent - Add 0.5 mL of solution 2 with 0.5mL of solution 3. to this mixture add 99 mL of solution 1. follow the same order as describes here and prepare the solution fresh

Solution 5: Folin-Ciocalteu reagent - Dilute the commercial reagent with an equal volume of distilled water on the day of use (This is a solution of sodium tungstate and sodium molybdate in phosphoric and hydrochloric acids)

Solution 6 : Standard protein -Bovine Serum Albumin - Dissolve 10mG of BSA in 10 mL of double distilled water

Solution 7 : Working standard - Makeup 1 mL of stock standard to 10mL. It gives 100µg/mL concentration

Other materials - Aluminium foil, Water bath, Test tubes, Standard measuring flask, Cuvette, Micropipette, Pipette.

Procedure

Pipette out various concentration of working standard solution into a series of test tubes and made up the volume to 0.2 mL with distilled water (10µl to 100µl).

To each test tube add 1 mL of the mixed reagent and mix thoroughly and allow to stand at room temperature for 10 min or longer.

Add 0.3mL of diluted Folin-Ciocolteau reagent rapidly and mix properly.

All tubes are incubated for 60 minutes.

OD of the standard and test solution was measured at 660nm and plot the standard graph. Run the blank.

The test protein sample is performed as like the standard solution and calculate the amount of protein present in the given sample.

Estimation of Protein-Lowry's Method

S. No	Volume of standard µl	Volume of water µl	Concentration of working sample	Mixed reagent (mL)	Folin's Reagent	OD-value (660nm)
1	10	190	10 µg	1	0.3	
2	20	180	20 µg	1	0.3	
3	30	170	30 µg	1	0.3	
4	40	160	40 µg	1	0.3	
5	50	150	50 µg	1	0.3	
6	60	140	60 µg	1	0.3	
7	70	130	70 µg	1	0.3	
8	80	120	80 µg	1	0.3	
9	90	110	90 µg	1	0.3	
10	100	100	100 µg	1	0.3	
Test sample 200 µl			Unknown	1	0.3	
Blank 200µl Water			Nil	1	0.3	000

Observation

Blue colour is noted and read using spectrophotometer

Result

Concentration of protein present in the given sample is ————µg/mL

I. ESTIMATION OF DNA BY DIPHENYLAMINE METHOD

Aim : To estimate the amount of DNA present in the sample.

Principle : When DNA is treated with diphenylamine under acid conditions, a blue compound is formed with a sharp absorption maximum at 595 nm. This reaction is given by 2-Deoxypentoses in general and is not specific for DNA (Adenine and guanine). In acid solution, the straight chain form of a deoxy pentose is converted to the highly reactive -hydroxylevulinic acid which reacts with diphenylamine to give a blue complex .

Materials Required

DNA standard solution : 5 mG of DNA as weighted accurately and make up to 25 mL with 5 mM sodium hydroxide.

Diphenylamine reagent : Dissolve 0.5g of pure diphenylamine in 48.7mL of glacial acetic acid. Add slowly 2.5 mL of concentrated sulphuric acid. Stir well and store for future use.

Other materials : Boiling water bath, Spectrophotometer, Aluminium foil, Water bath, Test tubes, Standard measuring flask, Cuvette, Micropipette, Pipette

Procedure

Pipette out different aliquots of standard DNA solution in the range of 0.05,0.1mL, 0.15 mL .....0.5mL with 0.05mL interval).

Make content of each tube to 1 mL using distilled water.

To the aliquots add 5 mL of diphenylamine reagent.

Close the test tubes with aluminium foil and keep it firm by rubber bands.

Heat on a boiling water bath for 10 min, cool and read the OD value at 595 nm.

Read the test and standards against water blank.

Note : The graph was drawn and amount of DNA in the given unknown solution was calculated. Perform estimation of unknown DNA as per previous procedure 2 to 6.

Estimation of DNA

S. No	Volume of standard solution in (mL)	Volume of water (mL)	Conc. of DNA in working sample (µg)	Volume of diphenyl-amine reagent (mL)	Incubation	OD-value (660nm)
01	0.05	0.95	10			
02	0.10	0.90	20			
03	0.15	0.85	30			
04	0.20	0.80	40			
05	0.25	0.75	50	5 mL	Boiling water bath for 10 min.	
06	0.30	0.70	60			
07	0.35	0.65	70			
08	0.40	0.60	80			
09	0.45	0.55	90			
10	0.50	0.50	100			
Unknown	0.2 mL	0.8 mL	Unknown			

Observation

Blue colour formation was observed.

Result

The amount of DNA present in the given sample is   ug/mL.

J. ESTIMATION OF RNA BY ORCINOL METHOD

Aim : To estimate the RNA from the given sample.

Principle : This is a general reaction for pentoses. Ribose Moieties of RNA form furfural when it is heated with concentrated hydrochloric acid. Orcinol reacts with the furfural in the presence of ferric chloride as a catalyst to give a brilliant green colour. Purine nucleotides only give any significant reaction.

Materials Required

Standard RNA solution : Dissolve 5 mG of RNA in 50 mL of Distilled water.

Orcinol solution : Dissolve 1 g of orcinol in 5 mL ethanol taken in a 100mL volumetric flask. Make up the content of the flask to 100mL using distilled water.

Concentrated hydrochloric acid

Ferric chloride solution (10%) : Dissolve 10g of ferric chloride in 100mL distilled water

Orcinol Reagent : Transfer 10mL of 10% ferric chloride solution to 390mL of conc. Hydrochloric acid. Mix this solution to 100 mL of orcinol solution. Continuously stir the mixture while adding and store it for future use.

Procedure

Different aliquots of standard RNA solution is taken and are make up to 3mL with distilled water (10µg to 100µg).

Add 3mL of orcinol reagent to all tubes.

Heat them on a boiling water bath for 10 minutes.

Observe for the development of colour.

After cooling them optical dencity is measured at 665 against blank.

Plot a graph between OD and the amount of RNA and from this standard curve.

Blank is prepared by using 3 mL of distilled water and 3 mL of orcinol and follow steps 3, 4 and 5.

2 mL of unknown sample is taken and made upto 3 mL and perform steps2, 3, 4 and 5.

Observation and Result

The amount of RNA present in the given sample is   /mL.

Estimation of RNA

S. No	Volume of standard solution in (mL)	Volume of water (mL)	Conc. of RNA in working solution (µg/mL)	Volume of Orcinol reagent	Incubation	OD-value (665nm)
01	0.1	2.9	10			
02	0.2	2.8	20			
03	0.3	2.7	30			
04	0.4	2.6	40			
05	0.5	2.5	50			
06	0.6	2.4	60	3 mL	Boiling water bath 10 min	
07	0.7	2.3	70			
08	0.8	2.2	80			
09	0.9	2.1	90			
10	1.0	2.0	100			
11 Unknown 2mL		1.0	Unknown			

K. ESTIMATION OF AMINOACID BY NINHYDRIN METHOD

Aim : To estimate free aminoacids from a test sample by ninhydrin method

Principle : Ninhydrin is one of a powerful oxidizing agent, decarboxylates the alpha aminoacids and yields an intensely coloured bluish product which is measured calorimetrically.

Nin hydrin + Aminoacid → Hydrindandin+Ammonia

Hydrindandin+Ammonia → Purple coloured product

Materials required

Ninhydrin: it is prepared by mixing solution A and B.

Solution A – Dissolve 0.8g stannous chloride in 500mL of 0.2M citrate buffer

Solution B- dissolve 20g of ninhydrin in 500mL of 2 methoxiethanol.

Diluent- mix equal volume of water and n propanol

Test tube, standard measuring flask, water bath, spectrophoto meter, 80% ethanol, conical flask.

Stock aminoacid: Dissolve 5mG of leucine in 50mL of Distilled water in a flask

Working standard: take 10mL stock and dilute to 100mL

Procedure

Standard graph :

Prepare working standard in a series of test tubes. Concentration of standard ranges from 10µg to100 µg (0.1mL to1mL).

Add one mL ninhydrin solution

Make up the volume to 2 mL with distilled water

Heat the tube in a boiling water bath for 20 minutes

Add 5mL of diluent and mix the content.

Read the intensity of the colour in spectrophotometer / calorimeter at 570nm after 15 minutes

Use 0.1mL of 80% ethanol along with other reagent as blank.

Test

Add one mL of ninhydrin to the 0.1mL of test sample.

Make up the volume to 2 mL with distilled water.

Heat the tube in a boiling water bath for 20 minutes.

Add 5mL of diluent and mix the content.

Read the intensity of the colour in spectrophotometer/calorimeter at 570nm after 15 minutes.

Use 0.1mL of 80% ethanol along with other reagent as blank.

Observation : Development of purple colour is noted.

Result : Draw standard graph and find quality of free aminoacid present in the sample.